Emerging Globalities and Civilizational Perspectives

Series Editor

Ino Rossi, Saint John's University, Great Neck, USA

This series documents the range of emerging globalities in the 21st century at the national, transnational and trans-civilizational levels of analysis. "Globality" refers to a global condition where people located at any point on Earth are aware of being part of the world as a whole---the world as a single interacting entity. Social interactions occur among actors belonging to different societies, different social strata and different cultural traditions so that the condition of "globality" is experienced in many different ways.

Examples of emerging globalities are social movements generated from the unfulfilled promises of neoliberalism and feelings of discrimination and marginalization of lower social strata; cultural otherization or the blaming of economic problems of certain geographical areas on a low level of cultural development; insecurities generated by technological risks, epidemics, and global terrorism; uncertainties generated by processes of transnational governance, outsourcing, unbalanced trade and massive migrations; biology-machine interfaces and impacts of non-human organisms and technologies on human consciousness and action; long-term threats of global warming, climate change and depletion of bio-diversity; increasing exploitation and marginalization of less industrialized regions.

We state that globalization entails encounters and often clashes among people and nations of different civilizational traditions. Hence, one of the exploratory questions of these volumes will be the extent to which negative or problematic globalities are reactions to failed promises and unrealized ideals of civilizational and national traditions and/or perhaps attempts to revive those traditions. Our notion of civilizational tradition takes inspiration from the classical works of Spengler and Toynbee, Benjamin Nelson, Vytautas Kavolis, Roland Robertson, Johann P. Arnason, Jeremy Smith, and others; a tradition which is in sharp contrast with the civilizationism recently promoted by authoritarian leaders with hegemonic ambitions. The volumes in this series aim to extend the inter-civilizational focus of classical civilizational thinkers from the analysis of the origins and development of civilizations to the fostering of contemporary inter-civilizational dialogues; the intent is to facilitate an international rapprochement in the contemporary atmosphere of global conflicts.

The volumes will reflect the diversity of theoretical perspectives and captures some of the novel thinking in social sciences, economics and humanities on intra- and inter-societal processes; the attention to novel thinking will extend to emerging policy formulations in dealing with threats, risks, insecurities and inequities and to strategic thinking for a sustainable global future. The historical perspective will also be an important component of analysis together with the avoidance of West-centric perspectives. The intended readership of this series is not just an academic audience but also policy decision-makers and the public at large; accessibility of language and clarity of discourse will be a key concern in the preparation of these volumes.

Martin Bohle • Boris Holzer • Leslie Sklair •
Fabienne Will

The Anthropocene Working Group and the Global Debate Around a New Geological Epoch

Springer

Martin Bohle
Institute for Globally Distributed Open Research and Education (IGDORE)
(Associated Fellow) Max Weber Centre for Advanced Cultural and Social Studies
Erfurt, Germany

Leslie Sklair
London School of Economics
London, UK

Boris Holzer
University of Konstanz
Konstanz, Baden-Württemberg, Germany

Fabienne Will
Research Institute for the History of Science and Technology
Deutsches Museum
Munich, Germany

ISSN 2731-0620 ISSN 2731-0639 (electronic)
Emerging Globalities and Civilizational Perspectives
ISBN 978-3-031-85174-2 ISBN 978-3-031-85175-9 (eBook)
https://doi.org/10.1007/978-3-031-85175-9

This Springer imprint is published by the registered company Springer Nature Switzerland AG
The registered company address is: Gewerbestrasse 11, 6330 Cham, Switzerland

Foreword

For a geologist, it is a strange feeling to be the object of study—and, in particular, to be the object of human-centred studies. After all, we represent the profession that, above all, looks out on our rocky planet from a position that is almost always cloaked in invisibility. Geology is one of the minor sciences, with a profile far below that of, say, physics or chemistry or biology; now rarely taught at schools, its institutions are commonly (like now) anxious about falling student numbers, and its public face is mostly dominated by dramatic tales of dinosaurs and volcanoes. More: geologists mostly live professionally among the near-infinite succession of decidedly non-human worlds that existed before our species arose: either at its surface among long-extinct life-forms amid unfamiliar geographies, or far below it, deciphering the inner workings of mountain belts and magma chambers. Many of us even (and I plead guilty here) were initially attracted to it because it was an escape from messy, human, socioeconomic realities.

And so how come one finds oneself here?—not only wholly immersed in study of the infinitely complex world of contemporary human impacts and their drivers, but also the object of attention of a study of the study itself? It typifies, perhaps, the looking-glass world of the Anthropocene, a concept that—often all too painfully—has turned the eternal verities of classical geology upside-down, in the epicentre of shock waves that have reached into distant scholarly disciplines to upend, say, the long-held distinction between nature and culture, as explored by Dipesh Chakrabarty, Bruno Latour and others. A further effect of this perturbation, though, has been to make those long-divided disciplines no longer so distant from each other—and that has made this book possible.

So where might one begin among the personal reflections that can be made on the Anthropocene Working Group? One is to go right back to the beginning of the geological analysis to note the contribution of the Geological Society of London's (GSL) Stratigraphy Commission. This was the body that made the first exploratory stratigraphic analysis of the Anthropocene, published in 2008 by *Geology Today* (Zalasiewicz et al. 2008), one of the Geological Society of America's journals, that led directly to the invitation from the Subcommission on Quaternary Stratigraphy (SQS) to form the Anthropocene Working Group (AWG).

As well as this link in developing the concept, and in contribution to the membership of the AWG (the Stratigraphy Commission provided 4 of the 16 original AWG members—Philip Gibbard, Andrew Kerr, Mark Williams and me, with Colin Waters joining shortly afterwards), the Stratigraphy Commission provided much of the template on which the AWG was based: a kind of working model, as it were, and for my part overtly so. This was not only because Mark and I, as the first Secretary and Chair, were much more familiar with the workings of the Stratigraphy Commission than of ICS working groups, but also because this national stratigraphic body had, at that time, developed a style and cohesive spirit that we valued and wished to maintain in this new Anthropocene venture (for which we had little other concrete idea, in those early days).

This Stratigraphy Commission's effectiveness had been much enhanced by my predecessor, Peter Rawson, a university academic who forged a close working relationship with Peter Allen, his counterpart at the British Geological Survey (BGS), and thus did much to end generations of mutual mistrust and academic territorial disputes between the university-based geological community and that of the national geological survey. This established a greater willingness of co-operation, which helped the work of the Stratigraphy Commission generally, and specifically as regards the Anthropocene as all three first Chairs and Secretaries (Mark Williams, Colin Waters—Secretary from 2012—and me) were attached both to the BGS and to universities at different stages of their careers.

Thus, when the invitation came to form the Anthropocene Working Group from Philip Gibbard, then Chair of the Subcommission on Quaternary Stratigraphy of the International Commission on Stratigraphy (ICS), the model used—though in part based on other working groups of the ICS, with which I had experience (the International Subcommission on Stratigraphic Nomenclature, of which I was a member from 2006 to 2012)—was more directly drawn from the GSL's Stratigraphy Commission for both Mark and me as we improvised the first 'executive' of the AWG. There were several reasons for this: the Stratigraphy Commission's focus then on the underlying science of any issue, rather than simply on technical correctness in nomenclature; a recognition that stratigraphy could provide insights of societal concern, as illustrated by statements on climate change; and a collegiate atmosphere that encouraged differences of opinion to be expressed both openly and amicably, as in the then long-running and sharp differences over whether or not to reinstate the formally obsolete but still widely used Tertiary Period (formally, this has been replaced by the Neogene and Paleogene periods) (Knox et al. 2012).

The focus on the science was helpful, indeed necessary, in the initial assembly of the membership of the AWG, because there was then so little technical information available as regards stratigraphical data on the Anthropocene. For, while Paul Crutzen, Will Steffen and their colleagues had already made a very strong case that the Earth System had changed from its Holocene state as regards ocean and atmospheric chemistry, climate change ecosystems and so on (Crutzen and Stoermer 2000; Crutzen 2002; Steffen et al. 2004), this was effectively building the case for an Anthropocene Epoch, an 'abstract' unit of time and process within the branch of stratigraphy formally termed geochronology (and that one needs to distinguish from its homonym, the determination of numerical age dates by methods such as radiometric dating). To make the Anthropocene geologically respectable—and definable—it needed to be as an Anthropocene Series: that is, as a material unit of strata that can be identified by distinctive and unique time-significant evidence such as fossils, and/or physical or chemical properties, and in particular through markers that can help recognise the boundary between it and Holocene deposits. There was no tradition of such a unit being recognised (before then, it had all been Holocene), and so the work had to begin from scratch, not least to answer the question of whether such a unit could be identified at all. As the AWG was beginning, there was no guarantee that such a unit existed—nor any real sense of how far through time it might extend, though Paul Crutzen's suggestion of the time since the Industrial Revolution made intuitive sense.

It was clear, too, that the membership of the AWG should extend beyond the discipline of stratigraphic geology, which pretty much exclusively made up all other ICS working groups. The Earth System Sciences (ESS) should naturally be represented, not least via Paul Crutzen and Will Steffen, who agreed to join—and showed a remarkable generosity of spirit, and a complete lack of academic territoriality or defensiveness, with respect to the concept that they, after all, had developed long before the geologists took notice. It seemed apt to represent also human and science history through such as John McNeill and Jacques Grinevald who had already deeply considered this question, and archaeology, the environmental sciences, oceanography, glaciology and so on. Other aspects came serendipitously, but in retrospect immensely usefully, as in the technosphere concept represented by Peter Haff, and the implications for international law in Davor Vidas. Stratigraphy, of course, included leading representatives of the Holocene and Quaternary—but also of much earlier episodes in Earth history, such as expertise in the Pliocene and Paleocene-Eocene Thermal Maximum, clear part-analogues for the Anthropocene. Membership evolved as new lines of evidence emerged to represent the growing realisation that environmental plastics were significant to the work, and industrial fly ash, and radionuclides. The work took place against a background of—and indeed was considerably enriched by, increasing wider interest that ranged into the social sciences, humanities and arts, exemplified by major exhibitions at Berlin's Haus der Kulturen der Welt (HKW) and the Deutsches Museum in Munich. Improvisation and serendipity played a large part in the assembly of the evidence, but by and large it worked out.

How to go about the work? For pretty well all other units of the Geological Time Scale (GTS), formalisation represents a tidying, a precise and agreed definition, of units which, for the most part, have been used for the last two centuries or so: units which have already been widely studied, units around which communities of stratigraphers have coalesced, units for which the most representative stratal successions have already been identified. None of this was the case for the Anthropocene, a unit with no geological history of study, and no dedicated geological community. The AWG found itself in the position of becoming this community, to explore, for the first time, whether the Anthropocene possessed scientific coherence, consistent with identity as a geological time unit.

The groundwork was done by the process that was to link the AWG members and create a kind of strong and cohesive 'AWG spirit', which persists to the present day, even after its transformation, now, to be an independent body. Essentially, a set of studies— various meta-analyses for the most part (e.g. Williams et al. 2011; Waters et al. 2014; Zalasiewicz et al. 2019b; and references therein)—was embarked upon to assemble and discuss stratigraphic evidence for the Anthropocene as a potential 'Series' of the chronostratigraphic scale, and not just as the 'Epoch' of time and process that was conceptualised by Crutzen, Steffen and their ESS colleagues. The work was improvised rather than rigorously planned, one study often showing the potential (or not) for another. Given the absence of funding for almost all of the AWG work (prior to the remarkable securing of funding by the HKW for the Global boundary Stratotype Section and Point (GSSP) sample analyses), the work was done mainly in moments of spare time (evenings and weekends figured all too often), with communication generally by email. AWG members were invited to join in on any current project, and they did so (or not) depending on their relevant expertise, enthusiasm for a particular topic, and availability. The only reward for their (considerable) work was the work itself, and part-authorship of whatever paper, thematic volume or book emerged—and here at least some of the publications attracted wider attention because of the general wider and public interest in the Anthropocene. There was scholarly interest too, so a number of AWG papers have turned out to be well cited. Studies were initiated by what one might loosely call the AWG executive, and also by individual AWG members, and most were circulated around the whole group for comment/invitation to take part prior to submission. One set of studies was generated by those within the AWG opposed to formalisation, although those were usually only seen by the rest of the AWG after their publication; nonetheless, they formed useful material for consideration and response. By one means or another, the wheels kept turning (and, indeed, turn still).

Sometime around 2015, the work slowly shifted in focus. By then, it had become clear that sufficient stratigraphic evidence existed to make an Anthropocene Series—the material unit of chronostratigraphy—defensible. Two major factors were in play here. Firstly, the realisation that the mid-20th century 'Great Acceleration' (Steffen et al. 2007; Steffen et al. 2015), rather than the Industrial Revolution of the late 18th or 19th century as originally suggested by Crutzen, was the key threshold in passing from a Holocene Earth System to an Anthropocene one. Secondly, that middle part of the second decade of the 21st century saw a number of independent studies published that recognised new stratigraphic proxy signals (and/or new patterns in established stratigraphic signals) that clearly translated a number of Great Acceleration phenomena into geological terms. Among these was the growing realisation that plastics were not only localised within the litter of urban areas, but were being carried by currents of air and water—particularly as microplastics—to virtually all parts of the globe, commonly in sufficient abundance to act as an Anthropocene marker in sedimentary layers (Zalasiewicz et al. 2016; and references therein). Another key marker emerged with the demonstration that the 'spherical carbonaceous particles' (SPCs) of industrial fly ash could also be used as a global marker, with the mid-20th century seeing their entry into strata in regions that industrialised relatively late, and showing a sharp 'Great Acceleration' upturn in regions (such as the UK) with a longer industrial history (Rose 2015; see also Swindles et al. 2015). These signals added to those earlier recognised, such as of persistent organic pollutants like pesticides (Muir and Rose 2007) and of artificial radionuclides disseminated by atmospheric bomb tests and civil nuclear accidents (Waters et al. 2015). Even inherently complex biological/palaeontological signals were found to be useable on the very brief time scale of the Anthropocene (Barnosky 2014; Wilkinson et al. 2014). The case for a sharply distinguishable Great Acceleration-based Anthropocene became, then, considerably stronger (Waters et al. 2016). This increasingly strong evidence base played a considerable part in the large majority in favour of a formalised Anthropocene that emerged in the first, informal, series of votes of the Anthropocene Working Group, held in 2016. The votes followed notable discussions at one of the few in-person AWG meetings, that was organised by Davor Vidas in Oslo. This included useful SQS/ICS perspective (both on ICS procedures, and regarding insights into resistance to the Anthropocene concept in ICS circles) via the presence of Martin Head, then SQS Chair, who was from then to contribute substantially to the AWG's work. These votes were finalised for announcement at the International Geological Congress (IGC) in Capetown that same year (Zalasiewicz et al. 2017a).

Significantly, neither this strengthened evidence base, nor further evidence subsequently gathered, did anything to diminish the outright opposition to the Anthropocene from a minority of AWG members and their colleagues (e.g. Gibbard and Walker 2014; Ruddiman 2018). This suggested that this opposition and that of others in the ICS—the strong opposition of the highly influential ICS Chair, Stanley Finney (Finney 2014; Finney and Edwards 2016) was a significant factor—even when responded to and countered (e.g. Zalasiewicz et al. 2017b; Zalasiewicz et al. 2019a) was not based on the amount and quality of stratigraphic evidence. Rather, it seemed to reflect more deep-seated aspects of the chronostratigraphically proposed Anthropocene (such as its brevity, and the fact that it does not include all anthropogenic impacts, but only those instrumental in moving the Earth System away from Holocene conditions).

The informal vote for the 2016 IGC set the stage for work towards a formal proposal. It was not straightforward to work out how to go about this, not least because the proposal was to be built around a GSSP (or stratotype, or 'golden spike', that is, a physical reference point in strata), which is the universally used means of defining a chronostratigraphic boundary in the later part of the Geological Time Scale (i.e. a little more than the last half billion years, the time interval in which fossils can be commonly used as stratigraphic indicators). An earlier suggestion that the Anthropocene boundary might be based on a simple numerical date (as is the case, still, for most units of the Precambrian) would have been more straightforward (Zalasiewicz et al. 2015) but seemed not to find favour with the ICS, and so the 'classical' GSSP route was chosen.

But where to start? For pretty well all other epochs of the GTS, there has been a century or more of research on the strata, and research communities encompassing several generations have built up around them, with many detailed research studies on rock successions from which a candidate stratotype might be chosen and, if necessary, analysed in further detail to precisely constrain the optimum level (the appearance of a particular fossil species, say, or the position of a distinctive chemical change) of the 'golden spike' level. Based on this long research history, there is usually considerable agreement between the specialists involved on both the best level and the best location for the spike—or, if not, at least any opposing positions have been well prepared and the arguments for and against well rehearsed. And, there is generally universal agreement that the unit to be defined should exist at all.

Little or none of this research history was in place for the Anthropocene, not least because its chosen beginning had just been decided—and, of course, its very existence was, and continued to be, strongly contested, irrespective of the detail of the stratigraphic data.

Furthermore, for this exercise, the AWG, uniquely multidisciplinary from the outset, had to be divided into two: a group of those who were specialist stratigraphers, and who could vote on any technical question to do with a proposed GSSP, and a group representing other disciplines, who continued to represent the wider AWG—and who indeed could continue to comment on stratigraphic matters—but who would not be included in the voting process. This process in the event took place amicably and without friction, largely by members self-identifying with one or other of these newly formed AWG divisions.

Finding and, more importantly, analysing candidate GSSP sections was initially more of a conundrum, partly because of the relatively unexplored nature of Anthropocene stratigraphy, and more importantly because the kind and number of analyses that were needed—of microplastic and fly ash content, of the isotopic compositions of carbon and nitrogen, of plutonium content, of different persistent organic pollutants and of much more—were very expensive, especially to a scientific group that had essentially been wholly unfunded for most of its work until then. It seemed, then, that the process would be very slow, to find promising locations, build scientific teams to work on them and somehow beg, steal or borrow the necessary funding. Even for 'established' geological units, it is not easy to obtain funding for GSSP work, and that is why it notoriously has taken decades—often several decades—to formalise many boundaries.

This problem was solved, remarkably and out of the blue, by the achievement of Bernd Scherer, Director of Berlin's HKW—who had both organised the first major exhibition around the Anthropocene and had maintained consistent interest in and support for this new concept (and for the AWG)—in obtaining, directly from the German government, 800,000 euros for the GSSP work, to be used over three years. Miraculous, but it brought its own problems in how to quickly find candidate locations, assemble research teams for each, devise a consistent analytical framework, handle the complex funding arrangements—and co-ordinate with the HKW's own substantial programme of arts and humanities-based activities based around the GSSP project. This was the very large task of Colin Waters and Simon Turner, by then AWG Chair and Secretary, respectively. (I had stepped down as Chair, acceding to a request to stand for Chair of the SQS, for better or worse, not least to try to provide a fair hearing for the by-now strongly controversial Anthropocene). The

GSSP task was, too, made substantially more difficult by the COVID pandemic and ensuing severe restrictions, as regards both fieldwork and laboratory analysis.

It was a task that they carried out with very considerable success. The twelve locations were partly sourced more or less from within the AWG, partly by building on projects in which AWG members were involved in, such as those at Searsville Reservoir, San Francisco Bay and the Baltic Sea, and partly by linking with existing, independent research projects such as those at Beppu Bay, Sihailongwan Lake, the Śnieżka peat and Antarctica. This allowed the funding to stretch further (the grant was generous, but the analyses involved were often very expensive).

This project did not only serve to provide, ultimately, a selected candidate GSSP but, perhaps more importantly, provided a large, integrated dataset that finally, unequivocally showed that the Anthropocene has all the qualities of an effective, clearly recognisable and precisely definable (and widely traceable) chronostratigraphic unit (that is, as a potential Series of strata). As with establishing, earlier, that the Anthropocene was geologically real, the outcome of this exercise could not be taken for granted, because it was the first one of its kind. Would these specific datasets be concordant, to further support a formalised Anthropocene? Or, at this very detailed scale of analysis—down to annual level, more or less unprecedented in stratigraphic correlation studies— would the data stubbornly refuse to fall into line, to cast doubt on the practicality of the Anthropocene as a consistently traceable unit of geological strata, i.e. as a potential Anthropocene Series? Geologists, after all, are all too familiar with the many factors—such as vagaries in currents of wind and water at the time of deposition of sedimentary particles—that can affect the pattern of environmental proxy signals that they strata enclose.

As it happened, the data, as it came in (Waters et al. 2024), showed effectively consistent and correlatable patterns, and furthermore clearly confirmed that the mid-20th century level is the optimal one, reflecting the appearance of many new stratigraphic signals, which in turn reflects a global threshold in the production and dispersal of a wide range of anthropogenic pollutants and signals of anthropogenic Earth System perturbation. This is most extravagantly seen in the Beppu Bay research (Kuwae et al. 2023), in which an astonishing 99 separate signals were distinguished and measured, their appearance or increase being shown statistically to cluster around the early 1950s.

Choosing, though, both the precise level for the base of a proposed Anthropocene Series (and simultaneously beginning of an Anthropocene Epoch), and choosing which site should serve as the GSSP, was not easy—indeed, the decisions were excruciatingly difficult and intensely discussed, because of the high quality of the data. As regards level, should one choose the very first, albeit only regionally dispersed appearance of plutonium that followed the Trinity test in 1945?— or the global dissemination of this radionuclide following the inception of the more powerful H-bomb tests in late 1952—or the appearance of some other signal—or even choose 1950 as a convenient date (already used as a datum in radiocarbon dating) very close to these? In the end, the 1952 date was chosen, largely because of how clearly this upturn was expressed by plutonium isotopes, the 'primary marker' and traceable in many of the datasets.

And as to which location? Here, the AWG was faced with an embarrassment of riches. On their own, most could easily have served as effective GSSPs, so it was a case of choosing from among the most tightly analysed and stratigraphically 'expressive' ones, with there being little to choose between (and further intense discussion of) Crawford Lake, Sihailongwan Lake and Beppu Bay. After several closely contested votes, Crawford Lake came out just clear, but it was a knife-edge decision (and the other two, together with the Śnieżka section, were chosen as auxiliary stratotypes). It is a shame that there has to be a 'winner' here, for in reality all twelve serve to describe—to personify, if one is allowed a little anthropomorphism—the Anthropocene.

From this study, systematically published in the April 2023 issue of The Anthropocene Review (Waters et al. 2023), the AWG had to produce, to a tight deadline of end of October 2023 imposed by the ICS, the formal proposal for the SQS to consider, debate and vote on. This was also a challenge, given the amount of data to synthesise, not just that accumulated in the GSSP exercise, but all of that compiled previously to demonstrate the reality of the Anthropocene in general, for the proposal needed to address the existential question as well as describe and analyse the proposed 'golden spike' itself, and the other supporting sections too.

The work was carried out, too, in a context of strained relations between the AWG and the higher echelons of the ICS and International Union of Geological Sciences (IUGS) executives, stemming from the long-standing opposition of key members of those executives to a formalised Anthropocene *per se*. The 190-page proposal (Waters et al. 2024) was submitted to the SQS by the very tight deadline imposed, with a couple of hours to spare, as a functional if incomplete document (as some samples were still undergoing laboratory analysis, not least the highest-resolution plutonium samples from Crawford Lake).

Preparation of the proposal had taken place among continued, indeed intensifying, opposition to formalisation led by the senior figures in the ICS and IUGS. A specific feature of the opposition had emerged in 2022 as an attempt (ongoing still) to radically change the meaning of the term away from that as formulated by Crutzen and the AWG, to that of an informal and variably interpretable 'Anthropocene event', essentially encompassing all detectable human environmental impacts going back 50,000 years and more, and within which the Great Acceleration and its planetary effects could be rendered well-nigh invisible (e.g. Fig. 1 of both Gibbard et al. 2022a; Gibbard et al. 2022b; subsequently reproduced in Ellis 2024 and other publications). A good deal of AWG time was absorbed in assembling detailed grounds to refute this proposed conceptual metamorphosis of the meaning of the Anthropocene (Head et al. 2022a; Head et al. 2022b; Waters et al. 2022; Syvitski et al. 2020). These evidence-based refutations did nothing to prevent further reiterations of the 'event' suggestion, again suggesting that the body of stratigraphic evidence assembled by the AWG was of little relevance to the central question of whether an Anthropocene epoch should exist at all.

The formal discussion of the AWG proposal at SQS level and beyond was highly unsatisfactory. As SQS Chair, my intention was to ensure full and open discussion of the proposal, including interchange with the AWG and wider consultation with appropriate representation of relevant outside bodies. In 2014, Stanley Finney, then ICS Chair, had set out specific guidelines in a paper for an AWG volume:

> Accordingly, the task group proposal must not only explain and justify its recommendation, but also address all questions that could be raised against it. Furthermore, when a task group's proposal is considered by the voting members of the subcommission, there must be extended opportunity for discussion in open meetings, via the internet, or both, and opinions from geoscientists outside of the subcommission should also be welcome and given careful consideration. It is necessary that all points are presented and all points challenged and all questions raised and all positions stated in these discussions. In such a fashion, voting members become fully informed. The same procedure occurs again when a subcommission recommendation is approved and forwarded to the ICS voting members. Final ratification by the IUGS Executive Commission of a recommendation approved by the ICS is based primarily on the assurance that ICS policies and procedures were followed (Finney 2014, p. 23).

This did not happen. Discussion of the proposal was minimal (only one SQS member provided detailed and useful review, though even here there was no time for the suggestions made to be incorporated); there was no significant interchange with the AWG, and no opportunity for outside consultation. The process essentially took place in closed session, with pressure for a quick vote, directly against my wishes as Chair, and those of one of the two vice-chairs, Martin Head. The vote was called anyway, circumventing the careful process outlined by Finney (2014) and cutting off further possibility of discussion (as the Statutes demand that discussion ends a month before the vote deadline). Neither Martin nor I took part, for we had no wish to give legitimacy to what we regarded as a formally invalid and scientifically entirely inadequate procedure.

The result (rejection) was announced via a pre-prepared New York Times article published the morning after the deadline, on March 5 2024. No feedback or commentary was given, simply the raw vote numbers (12 against and 4 in favour of formalisation). Comments, in those news releases, from the ICS Secretary-General were that the Anthropocene was too short (an oft-made criticism, and oft-answered by the AWG), and from the IUGS Secretary-General was that the Anthropocene was largely a means of getting publicity (for such comments, answers are unnecessary).

As Chair, I wrote a report noting that the vote had taken place in direct contradiction to Chair's advice (and thus in contradiction to ICS Statutes); that it had taken place when the majority of those who voted were ineligible to vote, having served more than 12 years (in direct contradiction of the Statutes also, where 12 years is the maximum for eligibility: indeed, if one counts votes eligible according to ICS Statutes and its 12-year limit for voting membership, then it is 2 against and 3 in favour: a 60% majority for acceptance) most considerably more; and hence stating that the vote should be declared null and void.

The matter quickly went to a quickly called meeting of ICS subcommission Chairs (from which I was excluded) who, after just a few days, upheld the rejection of the Anthropocene proposal by SQS members, a decision then ratified soon after by a meeting of the IUGS.

A further complication: the AWG listed a number of issues stemming from ICS/IUGS executive behaviour that, over past months and even years, it felt had hindered or undermined its work, submitting this to the Geoethics Commission of the IUGS for its independent opinion in November 2023. The IUGS examined the issue and made a number of recommendations that it reported directly to (and only to) the IUGS President in early 2024. The AWG did not learn the findings until the IUGS President released them on the morning after the vote—though only to AWG and SQS members with instructions that this report should go no further, and indeed its results remain withheld from further release to this today, in further demonstration of the 'closed tribunal' nature of the process. Overall, though, the IUGS did not act on the Geoethics Commission's main findings, which largely upheld the AWG's case and recommended that the vote be suspended. One day, we trust, the full story will be released, as a remarkable addition to the history of science.

This episode illustrates and reflects the strong reactions to the Anthropocene, and tensions raised, in the heart of the community that initially started the formal process. It is not normal that the ongoing work and eventual findings of a stratigraphic task group is criticised from within the highest levels of the geological time scale bureaucracy (usually a task group is regarded as being the main source of necessary expertise for any problem being examined). The Anthropocene clearly touches nerves that more ancient strata do not reach.

At some stage there will be a resubmission of the proposal. This need not wait for ten years (this ruling is for a new boundary, to give it sufficient initial stability, at least; for the Anthropocene such a boundary has not been established). The proposal, after all, though rejected, has not yet received formal detailed scrutiny and feedback. But any resubmission will likely need to wait until the conditions are in place for it to be calmly and dispassionately discussed within a framework of open science.

This fallout from the vote will affect the Anthropocene, I suspect, rather less than it will affect the science of geology. For, the Anthropocene is still with us and is growing more pronounced each year. When Paul Crutzen launched the term and the idea in 2000, for instance, one could, for instance, describe climate change as still incipient, not having yet caught up with the steep rise in greenhouse gas levels. Now, climate change is most definitely with us (though it has some way, yet, to go to catch up fully). Biodiversity levels continue to fall and the spread of novel materials such as plastics around the world continues, impacting ever more strongly on the Earth System. There are now Anthropocene-focused institutions, such as the Center for Anthropocene Studies at KAIST University, Daejeon, Korea, the Center for Anthropocene History at Stockholm's KTH Royal Institute of Technology, Sweden and the new Geoanthropology Max Planck Institute (the first ever multidisciplinary MPI) in Jena—and these use the Crutzen's meaning of the term, as a *de facto* epoch. Paul Crutzen's inspired hypothesis has been proven beyond any reasonable doubt: that the Earth System has been sharply shifted away from the long-held, relatively stable conditions of the Holocene; that most of the resulting changes are to all intents and purposes irreversible (i.e. there is no going back to Holocene conditions, though if we are very lucky we might get towards conditions that are in part Holocene-like); and that the Anthropocene is represented (and here this is through the AWG's work) in the form of a distinctive and globally traceable unit of strata (an Anthropocene Series), which can be precisely defined.

Geology, by contrast, has thrown away an opportunity to become involved in, and take ownership of, one of the most pressing and widely relevant scientific issues of this time. This may be especially damaging, because geology (and particularly stratigraphy) is the science which, arguably more than any other, has contributed to the forces driving Anthropocene change, in so inventively, ingeniously and successfully discovering and anatomising the resources of coal, oil and gas that have, more than anything else, allowed and driven the Great Acceleration. For the first few decades of the Great Acceleration, geologists could hold this up as a shining example of the great practical benefits of the science, in powering the post-WWII world and its economic boom. Now that the climate consequences are growing ever more serious, this is ever more a difficult posture to maintain—and to use as a means to persuade the younger generation to become geologists. Using the Anthropocene as a symbol of understanding—and then perhaps as a means of healing—a damaged planet would, I think, have been a better option. Formally denying its reality seems, at best, unhelpful.

As for the Anthropocene Working Group, this continues, and remains highly active, and has published detailed reflection on the vote and accompanying process (e.g. Oreskes 2024; Barnosky and Hannibal 2024; Turner et al. 2024), as well as further explorations of Anthropocene process (e.g. Williams et al. 2024; Summerhayes et al. 2024). Its membership is being increased (modestly, for now, as its dynamic has worked well with a few tens of members, but may well not with a cast of thousands). It is now independent, and so has more freedom of action, though broadly keeping its analysis within a deep time framework seems sensible for now. Following the completion of the GSSP exercise, it has reverted to its original, essentially unfunded, status. This has its limitations, but also gives considerable freedom of thought and action—though appropriate funding opportunities will be pursued, as occasion arises and as seems appropriate.

The AWG still has much to do. The proposal, assembled with such skill and effort by the 12 groups involved, still needs to be published properly, rather than as online preprints (Waters et al. 2024). There will be work to develop a wider consensus around the understanding of the term, so that it retains a clear meaning. Stabilisation of its meaning, after all, was the main purpose of the formalisation effort, for geological time terms have great (if not always absolute) inherent stability, once agreed. The importance for the Geological Time Scale and for geologists is much less than for the wider community. And, if this *de facto* stabilisation proves sufficiently effective in practice (Zalasiewicz et al. 2024), then the need for formal geological, chronostratigraphic definition will diminish, and the geology will become increasingly less relevant to the process.

Meanwhile, the interest shown in both the Anthropocene concept and how it has been developed—as shown by this carefully assembled and insightful volume—continues regardless of its formal status. The reality of the phenomenon, after all, continues to grow year by year and impact on all of our lives and indeed the lives of all the Earth's living organisms. Time will tell how it, and the efforts made to study it, will develop.

Acknowledgements I thank Fabienne Will and Martin Head for useful comments on the ms of this Foreword.

References

Barnosky, Anthony D. 2014. Palaeontological evidence for defining the Anthropocene. In *A Sratigraphical Basis for the Anthropocene?*, eds. Colin N. Waters, Jan A. Zalasiewicz, Mark W. Williams, Michael A. Ellis and Andrea M. Snelling, 149–165. Geological Society of London Special Publications, Bd. 395. London.

Barnosky, Anthony D. and Mary Ellen Hannibal. 2024. Despite Official Vote, the Evidence of the Anthropocene Is Clear. *Yale e360.* 2 April 2024. https://e360.yale.edu/features/anthropocene-denied. Last access: 13. August 2024.

Crutzen, Paul J. and Eugene F. Stoermer. 2000. The "Anthropocene". *Global Change Newsletter* 41: 17–18.

Crutzen, Paul J. 2002. Geology of mankind. *Nature* 415 (6867): 23. doi: 10.1038/415023a.

Ellis, Erle C. 2024. The Anthropocene is not an epoch – but the age of humans is most definitely underway. *The Conversation.* 5 März 2024. https://theconversation.com/the-anthropocene-is-not-an-epoch-but-the-age-of-humans-is-most-definitely-underway-224495. Last access: 13. August 2024.

Finney, Stanley C. 2014. The 'Anthropocene' as a Ratified Unit in the ICS International Chronostratigraphic Chart: Fundamental Issues that Must be Addressed by the Task Group. In *A Sratigraphical Basis for the Anthropocene?*, eds. Colin N. Waters, Jan A. Zalasiewicz, Mark W. Williams, Michael A. Ellis and Andrea M. Snelling, 23–28. Geological Society of London Special Publications, Bd. 395. London.

Finney, Stanley C. and Lucy E. Edwards. 2016. The "Anthropocene" epoch: Scientific decision or political statement? *GSA Today* 26 (3): 4–10. doi: 10.1130/GSATG270A.1.

Gibbard, Philip L. and Michael J. C. Walker. 2014. The term 'Anthropocene' in the context of formal geological classification. In *A Sratigraphical Basis for the Anthropocene?*, eds. Colin N. Waters, Jan A. Zalasiewicz, Mark W. Williams, Michael A. Ellis and Andrea M. Snelling, 29–37. Geological Society of London Special Publications, Bd. 395. London.

Gibbard, Philip L., Andrew M. Bauer, Matthew Edgeworth, William F. Ruddiman, Jacquelyn L. Gill, Dorothy J. Merritts, Stanley C. Finney, Lucy E. Edwards, Michael J. C. Walker, Mark Maslin and Erle C. Ellis. 2022a. A practical solution: the Anthropocene is a geological event, not a formal epoch. *Episodes* 45 (4): 349–357. doi: 10.18814/epiiugs/2021/021029.

Gibbard, Philip L., Michael J. C. Walker, Andrew Bauer, Matthew Edgeworth, Lucy E. Edwards, Erle Ellis, Stanley Finney, Jacquelyn L. Gill, Mark Maslin, Dorothy

Merritts and William Ruddiman. 2022b. The Anthropocene as an Event, not an Epoch. *Journal of Quaternary Science* 37 (3): 395–399. doi: 10.1002/jqs.3416.

Head, Martin J., Will Steffen, David Fagerlind, Colin N. Waters, Clément Poirier, Jaia Syvitski, Jan A. Zalasiewicz, Anthony D. Barnosky, Alejandro Cearreta, Catherine Jeandel, Reinhold Leinfelder, John Robert McNeill, Neil L. Rose, Colin Summerhayes, Michael Wagreich and Jens Zinke. 2022a. The Great Acceleration is real and provides a quantitative basis for the proposed Anthropocene Series/Epoch. *Episodes* 45 (4): 359–376. doi: 10.18814/epiiugs/2021/021031.

Head, Martin J., Jan A. Zalasiewicz, Colin N. Waters, Simon D. Turner, Mark W. Williams, Anthony D. Barnosky, Will Steffen, Michael Wagreich, Peter K. Haff, Jaia Syvitski, Reinhold Leinfelder, Francine M. G. McCarthy, Neil L. Rose, Scott L. Wing, Zhisheng An, Alejandro Cearreta, Andrew B. Cundy, Ian J. Fairchild, Yongming Han, Juliana Assunção Ivar do Sul, Catherine Jeandel, John Robert McNeill and Colin P. Summerhayes. 2022b. The proposed Anthropocene Epoch/Series is underpinned by an extensive array of mid-20 th century stratigraphic event signals. *Journal of Quaternary Science* 37 (7): 1181–1187. doi: 10.1002/jqs.3467.

Head, Martin J., Jan A. Zalasiewicz, Colin N. Waters, Simon D. Turner, Mark Williams, Anthony D. Barnosky, Will Steffen, Michael Wagreich, Peter K. Haff, Jaia Syvitski, Ian J. Leinfelder, R., McCarthy, F.M.G., Rose, N.L., Wing, Scott, L., An, Z., Cearreta, A., Cundy, A.B., Fairchild, Yongming Han, Juliana Assunção Ivar do Sul, Catherine Jeandel, John R. McNeill and Colin P. Summerhayes. 2023. The Anthropocene is a prospective epoch/series, not a geological event. *Episodes* 46 (2): 229–23.

Knox, R.W.O'B., P. N. Pearson, T. L. Barry, D. J. Condon, J.C.W. Cope, A. S. Gale, P. L. Gibbard, A. C. Kerr, M. W. Hounslow, J. H. Powell, P. F. Rawson, A. G. Smith, C. N. Waters and J. Zalasiewicz. 2012. Examining the case for the use of the Tertiary as a formal period or informal unit. *Proceedings of the Geologists' Association* 123 (3): 390–393. doi: 10.1016/j.pgeola.2012.05.004.

Kuwae, Michinobu, Bruce P. Finney, Zhiyuan Shi, Aya Sakaguchi, Narumi Tsugeki, Takayuki Omori, Tetsuro Agusa, Yoshiaki Suzuki, Yusuke Yokoyama, Hirofumi Hinata, Yoshio Hatada, Jun Inoue, Kazumi Matsuoka, Misaki Shimada, Hikaru Takahara, Shin Takahashi, Daisuke Ueno, Atsuko Amano, Jun Tsutsumi, Masanobu Yamamoto, Keiji Takemura, Keitaro Yamada, Ken Ikehara, Tsuyoshi Haraguchi, Stephen Tims, Michaela Froehlich, Leslie Keith Fifield, Takahiro Aze, Kimikazu Sasa, Tsutomu Takahashi, Masumi Matsumura, Yukinori Tani, Peter R. Leavitt, Hideyuki Doi, Tomohisa Irino, Kazuyoshi Moriya, Akira Hayashida, Kotaro Hirose, Hidekazu Suzuki and Yoshiki Saito. 2023. Beppu Bay, Japan, as a candidate Global boundary Stratotype Section and Point for the Anthropocene series. *The Anthropocene Review* 10 (1): 49–86. doi: 10.1177/20530196221135077.

Muir, D.C.G. and Neil L. Rose. 2007. Persistent organic pollutants in the sediments of Lochnagar. In *Lochnagar: The Natural History of a Mountain Lake*, ed. Neil

L. Rose, 375–402. Developments in Paleoenvironmental Research Ser, v.12. Dordrecht: Springer.

Oreskes, Naomi. 2024. Does the Anthropocene Matter?: The definition of this new geological epoch conveys the all-encompassing nature of the challenge we face. *Scientific American* 330 (1): 89. doi: 10.1038/scientificamerican0124-89.

Rose, Neil L. 2015. Spheroidal carbonaceous fly ash particles provide a globally synchronous stratigraphic marker for the Anthropocene. *Environmental Science & Technology* 49 (7): 4155–4162. doi: 10.1021/acs.est.5b00543.

Ruddiman, William F. 2018. Three flaws in defining a formal 'Anthropocene'. *Progress in Physical Geography: Earth and Environment* 42 (4): 451–461. doi: 10.1177/0309133318783142.

Steffen, Will, Angelina Sanderson, Peter Tyson, Jill Jäger, Pamela Matson, Berrien Moore III, Frank A. Oldfield, Katherine Richardson, Hans Joachim Schellnhuber, B. L. Turner, II and Robert J. Wasson (eds.). 2004. *Global Change and the Earth System: A Planet Under Pressure.* Berlin, Heidelberg: Springer.

Steffen, Will, Paul J. Crutzen and John Robert McNeill. 2007. The Anthropocene: Are Humans Now Overwhelming the Great Forces of Nature? *Ambio* 36 (8): 614–621.

Steffen, Will, Wendy Broadgate, Lisa Deutsch, Owen Gaffney and Cornelia Ludwig. 2015. The trajectory of the Anthropocene: The Great Acceleration. *The Anthropocene Review* 2 (1): 81–98. doi: 10.1177/2053019614564785.

Summerhayes, Colin P., Jan A. Zalasiewicz, Martin J. Head, Jaia Syvitski, Anthony D. Barnosky, Alejandro Cearreta, Barbara Fiałkiewicz-Koziel, Jacques Grinevald, Reinholf Leinfelder, Francine M.G. McCarthy, John R. McNeill, Yoshiki Saito, Michael Wagreich, Colin N. Waters, Mark Williams and Jens Zinke. 2024. The duration of the Anthropocene epoch: A synthesis. *Global and Planetary Change* 242, 104568.

Swindles, Graeme T., Elizabeth Watson, T. Edward Turner, Jennifer M. Galloway, Thomas Hadlari, Jane Wheeler and Karen L. Bacon. 2015. Spheroidal carbonaceous particles are a defining stratigraphic marker for the Anthropocene. *Scientific Reports* 5: 10264. doi: 10.1038/srep10264.

Syvitski, Jaia, Colin N. Waters, John Day, John D. Milliman, Colin Summerhayes, Will Steffen, Jan A. Zalasiewicz, Alejandro Cearreta, Agnieszka Gałuszka, Irka Hajdas, Martin J. Head, Reinhold Leinfelder, J. R. McNeill, Clément Poirier, Neil L. Rose, William Shotyk, Michael Wagreich and Mark Williams. 2020. Extraordinary human energy consumption and resultant geological impacts beginning around 1950 CE initiated the proposed Anthropocene Epoch. *Communications Earth & Environment* 1 (32). doi: 10.1038/s43247-020-00029-y.

Turner, Simon, Colin N. Waters, Jan A. Zalasiewicz and Martin J. Head. 2024. What the Anthropocene's critics overlook – and why it really should be a new geological epoch. *The Conversation.* 12 March 2024. https://theconversation.com/what-the-anthropocenes-critics-overlook-and-why-it-really-should-be-a-new-geological-epoch-225493. Last access: 26 March 2024.

Waters, Colin N., Jan A. Zalasiewicz, Mark W. Williams, Michael A. Ellis and Andrea M. Snelling (eds.). 2014. *A Sratigraphical Basis for the Anthropocene?* Geological Society of London Special Publications, Bd. 395. London.

Waters, Colin N., James P. M. Syvitski, Agnieszka Galuszka, Gary J. Hancock, Jan A. Zalasiewicz, Alejandro Cearreta, Jacques Grinevald, Catherine Jeandel, John Robert McNeill, Colin P. Summerhayes and Anthony D. Barnosky. 2015. Can nuclear weapons fallout mark the beginning of the Anthropocene Epoch? *Bulletin of the Atomic Scientists* 71 (3): 46–57. doi: 10.1177/0096340215581357.

Waters, Colin N., Jan A. Zalasiewicz, Colin P. Summerhayes, Anthony D. Barnosky, Clément Poirier, Agnieszka Galuszka, Alejandro Cearreta, Matthew Edgeworth, Erle C. Ellis, Michael A. Ellis, Catherine Jeandel, Reinhold Leinfelder, John Robert McNeill, Daniel D. Richter, Will Steffen, James P. M. Syvitski, Davor Vidas, Michael Wagreich, Mark W. Williams, An Zhisheng, Jacques Grinevald, Eric O. Odada, Naomi Oreskes and Alexander P. Wolfe. 2016. The Anthropocene is functionally and stratigraphically distinct from the Holocene. *Science* 351 (6269): aad2622. doi: 10.1126/science.aad2622.

Waters, Colin N., Mark W. Williams, Jan A. Zalasiewicz, Simon D. Turner, Anthony D. Barnosky, Martin J. Head, Scott L. Wing, Michael Wagreich, Will Steffen, Colin P. Summerhayes, Andrew B. Cundy, Jens Zinke, Barbara Fiałkiewicz-Kozieł, Reinhold Leinfelder, Peter K. Haff, John Robert McNeill, Neil L. Rose, Irka Hajdas, Francine M.G. McCarthy, Alejandro Cearreta, Agnieszka Gałuszka, Jaia Syvitski, Yongming Han, Zhisheng An, Ian J. Fairchild, Juliana Assunção Ivar do Sul and Catherine Jeandel. 2022. Epochs, events and episodes: Marking the geological impact of humans. *Earth-Science Reviews* 234 (Supplement 1): 1-28. doi: 10.1016/j.earscirev.2022.104171.

Waters, Colin N., Simon D. Turner, Jan A. Zalasiewicz and Martin J. Head. 2023. Candidate sites and other reference sections for the Global boundary Stratotype Section and Point of the Anthropocene series. *The Anthropocene Review* 10 (1): 3–24. doi: 10.1177/20530196221136422.

Waters, Colin N., Simon Turner, Zhisheng An, Anthony Barnosky, Alejandro Cearreta, Andrew Cundy, Ian Fairchild, Barbara Fiałkiewicz-Kozieł, Agnieszka Gałuszka, Jacques Grinevald, Irka Hajdas, Yongming Han, Martin J. Head, Juliana Assunção Ivar do Sul, Catherine Jeandel, Reinhold Leinfelder, Francine McCarthy, John McNeill, Eric Odada, Naomi Oreskes, Clément Poirier, Daniel Richter, Neil Rose, Yoshiki Saito, William Shotyk, Colin Summerhayes, Jaia Syvitski, Davor Vidas, Michael Wagreich, Mark Williams, Scott Wing, Jan A. Zalasiewicz und Jens Zinke. 2024. *Executive Summary: The Anthropocene Epoch and Crawfordian Age: proposals by the Anthropocene Working Group. Submitted to the ICS Subcommission on Quaternary Stratigraphy on October 31st, 2023.* https://eartharxiv.org/repository/view/6853/. Last access: 27. März 2024.

Wilkinson, Ian, Clément Poirier, Martin J. Head, C. D. Sayer and J. Tibby. 2014. Microbiotic signatures of the Anthropocene in marginal marine and freshwater palaeoenvironments. In *A Sratigraphical Basis for the Anthropocene?*, eds. Colin N. Waters, Jan A. Zalasiewicz, Mark W. Williams, Michael A. Ellis and Andrea

M. Snelling, 185–219. Geological Society of London Special Publications, Bd. 395. London.

Williams, Mark W., Jan A. Zalasiewicz, Alan Haywood and Michael A. Ellis (eds.). 2011. *The Anthropocene: a new epoch of geological time?*, Bd. 369. London.

Williams, Mark, Reinhold Leinfelder, Anthony D. Barnosky, Martin J. Head, Francine M. G.

McCarthy, Alejandro Cearreta, Stephen Himson, Rachael Holmes, Colin N. Waters, Jan A. Zalasiewicz, Simon Turner, Mary McGann, Elizabeth A. Hadly, M. Allison Stegner, Paul Michael Pilkington, Jérôme Kaiser, Juan Carlos Berrio, Ian P. Wilkinson, Jens Zinke and Kristine L. Delong. 2022. Planetary-scale change to the biosphere signalled by global species translocations can be used to identify the Anthropocene. *Palaeontology* 65, e12618.

Williams, Mark, Jan A. Zalasiewicz, Anthony D. Barnosky, Reinhold Leinfelder, Martin J. Head, Colin N. Waters, Francine M.G. McCarthy, Alejandro Cearreta, David C. Aldridge, Mary McGann, Paul B. Hamilton, Colin P. Summerhayes, Jaia Syvitski, Jens Zinke, Andrew B. Cundy, Barbara Fiałkiewicz-Kozieł, J. R. McNeill, Michinobu Kuwae, Neil L. Rose, Simon D. Turner, Yoshiki Saito, Michael Wagreich, M. Allison Stegner, Moriaki Yasuhara, Yongming Han, Amy Wrisdale, Rachael Holmes and Juan Carlos Berrio. 2024. Palaeontological signatures of the Anthropocene are distinct from those of previous epochs. *Earth-Science Reviews* 255: 104844. doi: 10.1016/j.earscirev.2024.104844.

Zalasiewicz, Jan A., Mark W. Williams, Alan Smith, Tiffany L. Barry, Angela L. Coe, Paul R. Bown, Patrick Brenchley, David Cantrill, Andrew S. Gale, Philip L. Gibbard, F. John Gregory, Mark W. Hounslow, Andrew C. Kerr, Paul Pearson, Robert Knox, John H. Powell, Colin N. Waters, John Marshall, Michael Oates, Peter F. Rawson and Philip Stone. 2008. Are we now living in the Anthropocene? *GSA Today* 18 (2): 4–8. doi: 10.1130/GSAT01802A.1.

Zalasiewicz, Jan A., Colin N. Waters, Mark W. Williams, Anthony D. Barnosky, Alejandro Cearreta, Paul J. Crutzen, Erle C. Ellis, Michael A. Ellis, Ian J. Fairchild, Jacques Grinevald, Peter K. Haff, Irka Hajdas, Reinhold Leinfelder, John Robert McNeill, Eric O. Odada, Clément Poirier, Daniel D. Richter, Will Steffen, Colin P. Summerhayes, James P. M. Syvitski, Davor Vidas, Michael Wagreich, Scott L. Wing, Alexander P. Wolfe, Zhisheng An and Naomi Oreskes. 2015. When did the Anthropocene begin?: A mid-twentieth century boundary level is stratigraphically optimal. *Quaternary International* 383: 196–203. doi: 10.1016/j.quaint.2014.11.045.

Zalasiewicz, Jan A., Colin N. Waters, Juliana Assunção Ivar do Sul, Patricia L. Corcoran, Anthony D. Barnosky, Alejandro Cearreta, Matthew Edgeworth, Agnieszka Galuszka, Catherine Jeandel, Reinhold Leinfelder, John Robert McNeill, Will Steffen, Colin P. Summerhayes, Michael Wagreich, Mark W. Williams, Alexander P. Wolfe and Yasmin Yonan. 2016. The geological cycle of plastics and their use as a stratigraphic indicator of the Anthropocene. *Anthropocene* 13: 4–17. doi: 10.1016/j.ancene.2016.01.002.

Zalasiewicz, Jan A., Colin N. Waters, Colin P. Summerhayes, Alexander P. Wolfe, Anthony D. Barnosky, Alejandro Cearreta, Paul J. Crutzen, Erle C. Ellis, Ian

J. Fairchild, Agnieszka Galuszka, Peter K. Haff, Irka Hajdas, Martin J. Head, Juliana Assunção Ivar do Sul, Catherine Jeandel, Reinhold Leinfelder, John Robert McNeill, Cath Neal, Eric O. Odada, Naomi Oreskes, Will Steffen, James P. M. Syvitski, Davor Vidas, Michael Wagreich and Mark W. Williams. 2017a. The Working Group on the Anthropocene: Summary of evidence and interim recommendations. *Anthropocene* 19: 55–60. doi: 10.1016/j.ancene.2017.09.001.

Zalasiewicz, Jan A., Colin N. Waters, Alexander P. Wolfe, Anthony D. Barnosky, Alejandro Cearreta, Matthew Edgeworth, Erle C. Ellis, Ian J. Fairchild, Felix M. Gradstein, Jacques Grinevald, Peter K. Haff, Martin J. Head, Juliana Assunção Ivar do Sul, Catherine Jeandel, Reinhold Leinfelder, John Robert McNeill, Naomi Oreskes, Clément Poirier, Andrew C. Revkin, Daniel deB. Richter, Will Steffen, Colin P. Summerhayes, James P. M. Syvitski, Davor Vidas, Michael Wagreich, Scott L. Wing and Mark W. Williams. 2017b. Making the case for a formal Anthropocene Epoch: An analysis of ongoing critiques. *Newsletters on Stratigraphy* 50 (2): 205–226. doi: 10.1127/nos/2017/0385.

Zalasiewicz, Jan A., Colin N. Waters, Martin J. Head, Clément Poirier, Colin P. Summerhayes, Reinhold Leinfelder, Jacques Grinevald, Will Steffen, James P. M. Syvitski, Peter K. Haff, John Robert McNeill, Michael Wagreich, Ian J. Fairchild, Daniel D. Richter, Davor Vidas, Mark W. Williams and Anthony D. Barnosky. 2019a. A formal Anthropocene is compatible with but distinct from its diachronous anthropogenic counterparts: a response to W. F. Ruddiman's "three-flaws in defining a formal Anthropocene". *Progress in Physical Geography* 43 (3): 319–333.

Zalasiewicz, Jan A., Colin N. Waters, Mark W. Williams and Colin P. Summerhayes (eds.). 2019b. *The Anthropocene as a Geological Time Unit. A Guide to the Scientific Evidence and Current Debate.* Cambridge: Cambridge University Press.

Zalasiewicz, Jan A., Julia A. Thomas, Colin N. Waters, Simon T. Turner, Martin J. Head and co-signatories (2024) What should the Anthropocene mean? *Nature* 632:980–984. https://doi.org/10.1038/d41586-024-02712-y

Emeritus Professor of Palaeobiology
at the University of Leicester
Leicester, UK

Jan Zalasiewicz

Preface

In times of planetary-scale anthropogenic change, natural and social sciences are confronted with whether human activities impacting the planet have ushered a new epoch. Many have followed the late Nobel Prize laureate Paul J. Crutzen (03. 12. 1933 - 21.01. 2021) and the freshwater biologist Eugene F. Stoermer (07. 04.1934 – 12. 02. 2012) in using the term “Anthropocene” for this, including the geological meaning. The International Union of Geological Sciences (IUGS), representing the international geological community, established the Anthropocene Working Group (AWG) in 2009 amid nascent debates about meaning and appropriate naming. The IUGS gave the AWG a limited geological mandate that clashes with broader concepts of an Anthropocene. This book inspects the effects of this ‘tension by design’.

The AWG took seven years to agree on proposing a novel geological epoch at the end of the Geological Time Scale, the Anthropocene. Received in 2016 at the occasion of the 35th International Geological Congress (IGC), the IUGS remained unsatisfied and reinstalled the AWG. The AWG submitted an elaborated proposal in October 2023, which the IUGS rejected in March 2024, some months before the 37th IGC. The book covers the entire period of existence of the AWG—as a technical body in the IUGS system—describing and interpreting the timeline of events, including the stalemate ruled by the IUGS in 2024. This book went to print when the AWG reconfigured as a scientific body *sui generis*.

During its existence, the AWG interacted intensively with scholars representing a wide range of natural and social sciences, using geological methodologies to formulate its proposal, being aware of broader debates about the notion of an Anthropocene. Specific features of this endeavour have been studied previously. This book looks at the dynamic of the entire process amid the globalised landscapes of contemporary sciences.

The book's impetus came from Leslie Sklair, who, in 2021, considered that the AWG's activities might present an interesting case study for the sociology of science. He assembled a multidisciplinary research group, combining four distinct research traditions for a joint co-authorship. The book's cooperative nature is expressed in the alphabetic order of the authors' names.

Versmold, Germany	Martin Bohle
Konstanz, Germany	Boris Holzer
Munich, Germany	Leslie Sklair
London, UK	Fabienne Will

Acknowledgements

This book has been a long time in the making, given that it follows a moving target. The for-what-ever-un/fortunate decision of the IUGS in March 2024 to reject the AWG's concept of a geological epoch Anthropocene was the opportunity to pause the studies supporting this book and to honour the editors' patience. The authors would like to thank their colleagues for advice, co-workers for help and families and friends for tolerance, such as the authors' temporary absence-mindedness. The numerous figures illustrating conceptual networks of the AWG were made possible thanks to the skills and dedication of research assistants at the University of Konstanz, who also carried out part of the bibliographic work.

Competing Interests The authors have no competing interests to declare that are relevant to the content of this manuscript.

Contents

About the Authors

Martin Bohle is an independent 'third-age-researcher' cooperating (board of experts) with the International Association for Promoting Geoethics (IAPG, Rome) since 2015. He was trained as a physical oceanographer (University Kiel, Germany, 1980) and is a lifetime member of the European Geosciences Union and the American Geophysical Union. Recent publications: *Geo-societal Narratives.* (Springer International Publishing, Cham; 2021, editor with E. Marone), *Phronesis at the Human-Earth Nexus: Managed Retreat.* (Front Polit Sci 4:1–13., 2022, with E. Marone), *Crafting Attributes of a Geological Now.* (Earth Sci Syst Soc 3:1–13; 2023). Martin Bohle worked in operational, executive and advisory functions from 1991-2019 at the Directorate General for Research and Innovation (DG RTD) of the European Commission. Before these experiences in science management, Martin Bohle researched the dynamics of coastal seas and lakes at the École Polytechnique Fédéral de Lausanne (PhD 1986, Switzerland) and the University of Hamburg (Germany). As a retired EU official (2019), he is affiliated with the Institute for Globally Distributed Open Research and Education (IGDORE) and the Max Weger Centre for Advanced Cultural and Social Studies (Associated Fellow); ResearchGate: D-4508-2014, ORCID: 0000-0002-8794-5810.

Boris Holzer is Professor of General Sociology and Macrosociology at the University of Konstanz. He has written on globalisation and modernisation theory, social networks and conflicts between protest groups and transnational corporations. His current research focuses on political and economic sociology, network theory and world society. Recent book publications include: From Globalization to World Society (ed. with F. Kastner and T. Werron, Routledge 2015), Schlüsselwerke der Netzwerkforschung (ed. with Ch. Stegbauer, Springer VS 2019) and Politische Soziologie (2nd ed., Nomos 2020).

Leslie Sklair is Emeritus Professor of Sociology, London School of Economics. He has published widely in sociological theory and the sociology of development and has been a consultant to the United Nations Centre on Transnational Corporations. In the 1980s, he carried out field research on the developmental impacts of foreign investment in Ireland, Egypt and (more intensively) China and Mexico. His works provided the material basis for Sociology of the Global System (published 1991, second updated edition in 1995, translated into Portuguese, Spanish, Japanese, Persian and Korean). He held various visiting Professorships and lectured at universities and conferences in the UK, Europe, North, Central and South America, Egypt, China, Hong Kong, Korea, Australia and Jamaica.

Fabienne Will is environmental historian and historian of science and currently affiliated with the Research Institute for the History of Science and Technology at the Deutsches Museum in Munich. There she is part of larger research project, the Munich Science Communication Lab on Planetary Health (MSCL). Her current research focuses on the history of wicked problems, moving at the interstice of environmental history, history of science and the history of technology. Before, she was part of the Rachel Carson Center of Environment and Society at LMU Munich (RCC) and has written on evidence practices in the Anthropocene debate.

List of Figures

List of Tables

Chapter 1
Introduction: The Anthropocene Working Group: Mission and Contexts

The Anthropocene's status as a cultural concept is highly controversial. Likewise, developing a supporting geological concept is challenging. Since its conception in 2000, the notion of the Anthropocene, both as a (simple) geoscientific and as a (complicated) cultural concept, has become increasingly familiar in the global public sphere (Sklair 2021). The International Union of Geological Sciences (IUGS) established the Anthropocene Working Group (AWG) in 2009, which has contributed to transforming the broad idea of 'planetary-scale anthropogenic change' into a geological (chronostratigraphic) conception of the end of the Holocene (Zalasiewicz et al. 2023). Based on its mandate and topic, the AWG has been embedded in international academic, cultural, and bureaucratic environments, making it a microcosm of globalized science.

This book describes the AWG's work from 2009 to 2023, the lifespan as a working group of the International Commission on Stratigraphy (ICS), a constituent body of the International Union of Geological Sciences (IUGS), both overseeing the AWG's execution of its mandate, that is, seeking a possible chronostratigraphic foundation of the Anthropocene concept. The IUGS did not approve the AWG's affirmation of a novel geological epoch (Anthropocene) in March 2024 (IUGS 2024). Instead, it favors the concept of a geological event, i.e., of a cumulative articulation of cultural and geological features (Koster et al. 2024). Despite the IUGS' decision, the debate about the political, cultural, and philosophical attributes of the Anthropocene concept continues (Ansede 2024; De Jouvancourt 2024; Barnosky and Hannibal 2024). Looking beyond the debate in the natural sciences, this book also examines how social scientists, historians, environmental humanities scholars, and creative artists have responded, taking up the notion of a geological Anthropocene as a way to make sense of the planetary crisis, including the climate crisis and the future of human life on planet Earth. Addressing notions of a

This chapter is written by Martin Bohle, Boris Holzer, Leslie Sklair, Fabienne Will

M. Bohle et al., *The Anthropocene Working Group and the Global Debate Around a New Geological Epoch*, Emerging Globalities and Civilizational Perspectives, https://doi.org/10.1007/978-3-031-85175-9_1

geological *and* cultural Anthropocene, we attempt to understand why mainstream geology rejects the concept of a geological epoch Anthropocene (Waters 2023) in favor of the concept of a geological event (Walker et al. 2024).

1.1 The Anthropocene Working Group, Science, and the Public

The Anthropocene proposal, popularized at the dawn of the twenty-first century by Paul Crutzen (Benner et al. 2021), marks a significant turning point in recognizing anthropogenic impacts on a planetary scale. Its essence is the end of the Holocene, the geological period during which human societies developed. At the turn of the last century, only a few observers thought that the planetary-scale anthropogenic change heralded the inception of a new geological epoch. Within two decades, this initial conceptual leap was transposed into a proposal to amend the Geological Time Scale (GTS), formally made by the AWG. The AWG was a temporary, low-level working party of a constituent body of the IUGS. With the formal submission of the proposal[1] (October 31st, 2023) to its parent body, the AWG ceased to exist within the IUGS system. The fact that the AWG's proposal was rejected (IUGS 2024) at the inception level (Witze 2024a, 2024b) took many people outside the geoscientific community by surprise. Although this decision will not terminate the debate about the Anthropocene—both in the geosciences and beyond—it marks a preliminary conclusion of the debate about the geological interpretation and operationalization of the Anthropocene, which has spanned over 20 years.

The case of the AWG provides an example of how globalized science, interdisciplinary debates, and public discourse intersect on a topic that has slightly different meanings for each of them. From a geoscientific perspective, the Anthropocene presents a 'puzzle' to be solved through 'normal science' (Kuhn 1962): Can the idea of a 'new epoch' be operationalized and empirically validated using established concepts and methods? With the Geological Time Scale, geology offers a conceptual framework that could accommodate the Anthropocene idea—if appropriately operationalized and specified. From the perspective of other disciplines, however, the Anthropocene's meaning and potential are determined by entirely different conceptual frameworks, often implying less stringent procedures to assess its empirical validity but more latitude to include a variety of dimensions and determinants. For instance, the social sciences and the humanities are more inclined to align their concepts with the broader public understanding of the Anthropocene that stresses its relation to anthropogenic climate change.

[1] https://quaternary.stratigraphy.org/working-groups/ (accessed 20th February 2024; the site been reorganised after that date) "The Anthropocene Working Group (AWG) was established in June 2009 to examine the possibility of recognising an Anthropocene division either within the Holocene or separated from it. A formal proposal was submitted to the SQS on October 31, 2023, that the Anthropocene be accepted as a stage/epoch terminating the Holocene, with a GSSP in a core recovered from Crawford Lake, near Milton, Ontario, Canada that would also define the Crawfordian Stage/Age."

Consequently, they tend to regard the Anthropocene as a cypher for the global ecological crisis. Oscillating between these different contexts, the Anthropocene concept can be considered a communicative 'boundary object,' i.e. a concept that is "both plastic enough to adapt to local needs and the constraints of the several parties employing them, yet robust enough to maintain a common identity across sites" (Star and Griesemer 1989, p. 393). While the Anthropocene thus connects the scientific debate in the natural and social sciences with a broader public discourse, it also leads to friction and misunderstandings. For instance, conceptual clarification and empirical methods may be able to determine what the Anthropocene 'is' and when it 'began'—but in this process, the meaning of the term is not only specified but also redefined. While some may expect a formal endorsement by the AWG and a more precise conceptualization to give the Anthropocene more credibility and authority, the 'geological' Anthropocene could eventually be quite different from its common-sense interpretation. To prevent such a divorce of geological and other interpretations, the AWG sought to integrate various disciplinary and regional perspectives while acknowledging the epistemological and institutional constraints of the discipline and its organizational environment.

However, the inter- and transdisciplinary dimension of the Anthropocene debate gave rise to a latent (and intermittently manifest) conflict between disciplinary autonomy and societal relevance. While some geoscientists accepted the challenge of translating the societal concern with the Anthropocene into a consistent and scientifically fruitful concept that could, in turn, inform the public debate, others were wary about the potential overlap of scientific puzzle-solving and political problem-solving. Not only for geoscientists, scientific authority appears to rest on their practices being 'disinterested' in political and other societal matters (Bourdieu 1975). From this perspective, importing the relevance (if not the definition) of a problem from outside science may be regarded as challenging science's authority and autonomy. As we will show in the following chapters, this tension between geoscientific and other understandings of the Anthropocene accompanied the work of the AWG as both a stimulating and a problematic aspect of its position within the geoscientific community and in the broader public debate around the Anthropocene.

1.2 Structure of the Book

This book examines the AWG's journey from 2009 to 2023 (Chap. 2), the AWG's endeavors to transcend the disciplinary confines of geoscience (Chaps. 3 and 4), and the AWG's engagement with a diverse audience spanning academia and the public sphere (Chap. 5). The concluding Chap. 6 discusses the circumstances that led the IUGS to reject the AWG's proposal to fix the onset of the Anthropocene epoch in the mid-twentieth century.

Chap. 2 shows that the AWG was a central and dynamic force within the geoscientific community, capable of influencing academic discourse and public understanding of scientific concepts like the (geological) Anthropocene. The AWG's ability to navigate bureaucratic constraints, form strategic alliances, and engage with broader societal issues facilitated its significant impact within and beyond the geoscientific community.

1. Nature and Role: The AWG is portrayed as an autonomous group of academic peers focusing on a methodological issue within their discipline. It operated within the bureaucratic framework of the IUGS, the ICS, and the ICS's Subcommission on Quaternary Stratigraphy (SQS). Nevertheless, the AWG's relevance extended beyond geological confines due to its work's global and societal implications.
2. Challenges: The AWG struggled with its position within the bureaucratic structures of IUGS, which is not equipped to address issues of global or planetary societal relevance. These struggles included navigating a complex, multifaceted global environment, dealing with constraints imposed by these structures, and handling opportunities and external material contexts.
3. Resilience and Alliances: The AWG demonstrated intellectual resilience and learned to form strategic coalitions and mobilize external support. This ability has been crucial in its efforts to address and transcend the limitations posed by its traditional disciplinary boundaries, although insufficient to secure the fate of its proposal.

Chapter 3 investigates the position of the AWG in scientific discourse(s) about the Anthropocene based on its members' affiliations and publications. Current and former AWG members' disciplinary and institutional affiliations reflected its focus on geoscience and its aim to include other perspectives.

4. Diversity: Although most AWG members have a disciplinary background in geosciences (sensu *largo*), the interdisciplinarity of the Anthropocene question is reflected in the participation of scholars from other disciplines or with a broader "Earth System" perspective. Their geographical affiliations are primarily to academic institutions in North America and Western Europe, with the latter weighted to the United Kingdom.
5. Reputation: According to indicators of scholarly reputation, such as citations, the AWG members are well integrated into their respective fields and thus lend credibility and authority to the proceedings and decisions of the AWG.
6. Cooperation: The analysis of the scholarly output of AWG members highlights multiple facets of the Anthropocene concept. A text corpus of scientific publications mentioning the Anthropocene (from 2000 to 2020) corroborates the visibility of the AWG based on citation patterns involving AWG members, with some influential publications also bridging thematic and disciplinary clusters.

Chapter 4 describes the AWG through exploring and developing the (cultural) Anthropocene concept. The AWG is positioned as a catalyst for interdisciplinary collaboration, challenging traditional disciplinary boundaries and fostering a more integrated understanding of Earth's history and humanity's role within it. This feature has significant implications within the geoscientific community and the broader realms of humanities, social sciences, and public discourse.

7. The Anthropocene concept: The evolution of the Anthropocene reflects a multifaceted notion, not bound by a single perspective or discipline. This complexity and the blurring of traditional disciplinary boundaries underscore the provocative potential of the Anthropocene debate.
8. Integrating diverse scales: A significant focus is the combination of different temporal, spatial, and conceptual dimensions. The Anthropocene concept enables the linking of geological and historical times, creating new narratives and the idea of geo-history, and emphasizes the role of historians and humanities in integrating the planetary scale into historiographical studies. Assuming this role involves developing narratives encompassing deep geological time and historical events, which is crucial in translating geoscientific discourse to broader contexts.
9. Interdisciplinary collaboration: The AWG's work highlights the challenges and opportunities in bridging the gap between natural sciences, humanities, and social sciences. This multidisciplinary approach is essential for fully understanding and contextualizing the Anthropocene concept, particularly its ethical, political, social, and economic implications. The future trajectory of the Anthropocene debate suggests that it could lead to new lines of thought and narratives, potentially forming a new interdisciplinary field that transcends traditional academic boundaries.

Chapter 5 describes how the AWG has been instrumental in advancing Anthropocene concepts across various disciplines and societal spheres. Its extensive outreach, interdisciplinary approach, and engagement with complex debates demonstrated the group's significant impact and the ongoing evolution of the Anthropocene narrative.

10. Outreach Efforts: The AWG's outreach has been extensive, crossing traditional academic boundaries and engaging with various disciplines, media, and creative arts. This outreach enables the geological Anthropocene concept to resonate beyond natural sciences in broader societal and cultural arenas. AWG's outreach approach is akin to a business strategy, aiming to persuade and engage diverse audiences.
11. Debate and Controversy: The geological Anthropocene concept has sparked debate within the geological community, with some viewing it more as a "pop culture" phenomenon than a stratigraphic one. The AWG's work highlights the intersections between scientific/geological issues and broader political, social, and existential attributes associated with the Anthropocene concept. The group's commitment to interdisciplinarity has kept discussions open to these wider implications.
12. Future Directions: The future of the AWG and the Anthropocene concept is presented as an open question. The group's leadership is noted, suggesting that the term Anthropocene has been firmly established across scientific disciplines and languages, regardless of the resolution of nomenclature issues within geological sciences.

1.3 Mission Accomplished, or Mission Failed?

Through an interdisciplinary approach (Zalasiewicz et al. 2019), the AWG amalgamated Earth System Science (ESS), geological stratigraphy, and contemporary observational techniques to argue for ending the geological epoch Holocene and establishing the novel epoch Anthropocene. Compared to the initial suggestion of the onset of the Anthropocene during the Industrial Revolution (Crutzen 2002), the AWG suggested (Waters et al. 2016; Zalasiewicz et al. 2019; Witze 2023) to situate the Anthropocene's onset in the mid-twentieth century—a period of planetary system shift, notably characterized as the Great Acceleration (Waters et al. 2023). This innovative choice has facilitated a multifaceted Anthropocene narrative, including challenging views like "runaway sociocultural niche construction." (Ellis 2024) and invited contributions from social scientists, environmental humanities scholars, and creative artists (Thomas 2022).

Given the groundbreaking nature of the initial Anthropocene Proposal, the journey toward formally proposing ratification within the GTS could not be a straightforward path. Although, as aforementioned, defining the Anthropocene and deciding when it began appeared to be a matter of 'normal science,' certain features made that a thorny or 'provocative' issue for geoscience. On the one hand, the implication of recent anthropogenic influence runs counter geology's focus on long-term developments and past events beyond human interference. On the other hand, the many-faceted nature of the Anthropocene does not lend itself to a clear demarcation in both time and space. While the AWG chose well-established methods, such as marking the onset of an epoch through a 'golden spike,' its openness to interdisciplinary questions and concerns also hinted at the possibility of a more encompassing paradigmatic shift. For these and other reasons, some IUGS senior officers were known to be critical of the idea of an epoch Anthropocene, which contributed to the eventual rejection of the AWG's proposal. However, the way in which the proposal was aborted also shows a high determination to halt further intraorganizational discussions.

Whether and how the scientific discussion about a geological Anthropocene will proceed is speculative. Considering that the detailed AWG proposal is publicly available (Waters et al. 2024a, 2024b), confirming its scientific quality, there is good reason to expect continuing debates. Beyond geological terminology, the AWG's scientific work of nearly two decades has triggered a paradigm shift across geoscience, philosophical, and political domains, including calls to reevaluate established norms and policy strategies of the Holocene (Dryzek and Pickering 2018; Wallenhorst 2023; Rockström et al. 2023), which are needed to cope with the consequences of planetay-scale anthropogenic change.

The AWG's scientific 'post-normal' methodology, which integrates geological stratigraphy, Earth System Sciences, and communication practices, confirms a commitment to addressing the Anthropocene as a scientific problem and simultaneously emphasizes the need for a broad, transdisciplinary understanding of its implications. This approach highlights the Anthropocene debate's role as a bridge linking geological and historical timescales, i.e., developing narratives that integrate the planetary scale into historiographical studies. The AWG's public outreach and engagement efforts have propelled the Anthropocene concept beyond geology, stimulating debates addressing technical, political, social, and existential dimensions. This broad engagement underscores the AWG's essential role in developing a post-normal mode of science that brings together diverse academic disciplines to grasp the transformative impact of the Anthropocene concept.

The work and trajectory of the AWG, starting from the initial suggestion of Paul Crutzen to making a formal proposal, have highlighted the complex interdependencies among academic disciplines, the global public's perception of science, and the issue of planetary-scale anthropogenic change. Through a comprehensive analysis of the work of the AWG and its interdisciplinary and cultural contexts, the following chapters document the AWG's role in advancing the Anthropocene concept and a novel understanding of the role of science in a changing world.

References

Ansede, Manuel. 2024. *The Anthropocene war: The controversial vote that decided not to change the planet's geological epoch.* El Pais. https://english.elpais.com/science-tech/2024-03-08/the-anthropocene-war-the-controversial-vote-that-decided-not-to-change-the-planets-geological-epoch.html# (accessed 11th March 2024)

Barnosky, Anthony D., and Ellen Hannibal. 2024. Despite official vote, the evidence of the Anthropocene is clear. *Yale Environment* 360. https://e360.yale.edu/features/anthropocene-denied (accessed 5th Dec. 2024)

Benner, Susanne, Gregor Lax, Paul J. Crutzen, Ulrich Pöschl, Jos Lelieveld, and Hans Günter Brauch. 2021. *Paul J. Crutzen and the Anthropocene: A new epoch in Earth's history*. In *The Anthropocene: Politik—Economics—Society—Science*, ed. Susanne Benner, Gregor Lax, Paul J. Crutzen, Ulrich Pöschl, Jos Lelieveld, and Hans Günter Brauch, vol. 1. Cham: Springer International Publishing. https://doi.org/10.1007/978-3-030-82202-6.

Bourdieu, Pierre. 1975. The specificity of the scientific field and the social conditions of the progress of reason. *Social Science Information* 14:19. https://doi.org/10.1177/053901847501400602.

Crutzen, Paul J. 2002. Geology of mankind. *Nature* 415: 23. https://doi.org/10.1038/415023a

De Jouvancourt, Pierre. 2024. Qui a tué l'Anthropocène ? *Terrestres. https://www.terrestres.org/2024/04/08/qui-a-tue-lanthropocene/ (accessed 1st May 2024)*

Dryzek, John S., and Jonathan Pickering. 2018. *The politics of the Anthropocene*. Oxford: Oxford University Press. https://doi.org/10.1093/oso/9780198809616.001.0001.

Ellis, Erle C. 2024. The Anthropocene condition: Evolving through social–ecological transformations. *Philosophical Transactions of the Royal Society B: Biological Sciences* 379:20220255. https://doi.org/10.1098/rstb.2022.0255.

IUGS. 2024. *The anthropocene*. https://www.iugs.org/_files/ugd/f1fc07_40d1a7ed58de458c9f8f24de5e739663.pdf (accessed 22nd March 2024)

Koster, Emlyn, Philip L. Gibbard, and Mark Maslin. 2024. The Anthropocene event as a cultural zeitgeist in the earth-human ecosystem. *Journal of Geoethics and Social Geosciences* 01:1–41. https://doi.org/10.13127/jgsg-43.

Kuhn, Thomas S. 1962. *The structure of scientific revolutions*. The University of Chicago Press. International Encyclopedia of Unified Science. Vol. II.

Rockström, Johan, Joyeeta Gupta, Dahe Qin, Steven J. Lade, Jesse F. Abrams, Lauren S. Andersen, David I. Armstrong McKay, et al. 2023. Safe and just earth system boundaries. *Nature* 619:102. https://doi.org/10.1038/s41586-023-06083-8.

Sklair, Leslie. 2021. Anthropocene in the mass media: The big picture. In *The Anthropocene in global media: Neutralizing the risk,* ed. Leslie Sklair, 22–34. Abingdon: Routledge. https://doi.org/10.4324/9780429355202.

Star, Susan Leigh, and James R. Griesemer. 1989. Institutional ecology, `translations' and boundary objects: Amateurs and professionals in Berkeley's Museum of Vertebrate Zoology, 1907-39. *Social Studies of Science* 19:387–420. https://doi.org/10.1177/030631289019003001.

Thomas, Julia Adeney. 2022. In *Altered earth*, ed. Julia Adeney Thomas. Cambridge, UK: Cambridge University Press. https://doi.org/10.1017/9781009042369.

Walker, Michael J. C., Andrew M. Bauer, Matthew Edgeworth, Erle C. Ellis, Stanley C. Finney, Philip L. Gibbard, and Mark Maslin. 2024. The Anthropocene is best understood as an ongoing, intensifying, diachronous event. *Boreas* 53:1–3. https://doi.org/10.1111/bor.12636.

Wallenhorst, Nathanaël. 2023. *A critical theory for the Anthropocene. Anthropocene—Humanities and social sciences*. Cham: Springer International Publishing. https://doi.org/10.1007/978-3-031-37738-9.

Waters, Colin N. 2023. *Executive Summary The Anthropocene Epoch and Crawfordian Age: proposals by the Anthropocene Working Group*. https://eartharxiv.org/repository/view/6853/ (accessed 16 March 2024)

Waters, Colin N., Simon D. Turner, Jan Zalasiewicz, and Martin J. Head. 2023. Candidate sites and other reference sections for the global boundary Stratotype section and point of the Anthropocene series. *The Anthropocene Review* 205301962211364:3. https://doi.org/10.1177/20530196221136422.

Waters, Colin N., Jan Zalasiewicz, Colin Summerhayes, Anthony D. Barnosky, A. Clément Poirier, Alejandro Cearreta Gauszka, et al. 2016. The Anthropocene is functionally and stratigraphically distinct from the Holocene. *Science* 351:aad2622–aad2622. https://doi.org/10.1126/science.aad2622.

Witze, Alexandra. 2023. This quiet lake could mark the start of a new Anthropocene epoch. *Nature* 619:441–442. https://doi.org/10.1038/d41586-023-02234-z.

———. 2024a. Geologists reject the Anthropocene as Earth's new epoch—After 15 years of debate. *Nature* 627:249. https://doi.org/10.1038/d41586-024-00675-8.

———. 2024b. It's final: The Anthropocene is not an epoch, despite protest over vote. *Nature*. https://doi.org/10.1038/d41586-024-00868-1.

———. 2024b. *The Anthropocene Epoch and Crawfordian Age: proposals by the Anthropocene Working Group Part 1: Anthropocene Series/Epoch: stratigraphic context and justification of rank*. https://eartharxiv.org/repository/view/6954/ (accessed 17th April 2024)

Waters, Colin N., Simon Turner, An Zhisheng, Anthony Barnosky, Alejandro Cearreta, Andrew Cundy, Ian Fairchild, et al. 2024a. *The Anthropocene Epoch and Crawfordian Age: proposals by the Anthropocene Working Group Part 2: Descriptions of the proposed Crawford Lake GSSP and supporting SABSs*. https://eartharxiv.org/repository/view/6963/ (accessed 17th April 2024)

Zalasiewicz, Jan, Colin N. Waters, Mark Williams, and Colin Summerhayes. 2019. In *The Anthropocene as a geological time unit*, ed. Jan Zalasiewicz, Colin N. Waters, Mark Williams, and Colin P. Summerhayes. Cambridge University Press. https://doi.org/10.1017/9781108621359.

Zalasiewicz, Jan, Colin Waters, Simon Turner, Mark Williams, and Martin J. Head. 2023. Anthropocene working group. In *Handbook of the Anthropocene*, 315–321. Cham: Springer International Publishing. https://doi.org/10.1007/978-3-031-25910-4_48.

Chapter 2
Timelines and Entanglements

2.1 Introduction

To describe the geological 'present', Paul Crutzen proposed the term Anthropocene (Crutzen and Stoermer 2000) at a workshop of the International Geosphere-Biosphere Programme[1] (IGBP). It is unknown whether the workshop participants were conscious of the implied epistemic trespassing (Ballantyne 2019) and challenges (Benner et al. 2021). However, the term rapidly acquired multiple meanings (Zalasiewicz et al. 2021; Table 1, p.11,12); for example, in Earth System Sciences[2] (Steffen et al. 2020, Box 2 p. 60)

> "The Anthropocene as proposed in 2000 had two meanings. In a geological context, Crutzen proposed the Anthropocene as a new epoch to follow the Holocene in the Geological Time Scale (GTS). In an Earth System context, the Anthropocene was proposed as a very rapid trajectory away from the 11,700-year, relatively stable conditions of the Holocene. The two definitions, although not identical, have much in common."

The International Commission on Stratigraphy (ICS), a constituent body of the International Union of Geological Sciences (IUGS), waited for the better part of a decade to establish the 'Working Group on the Anthropocene' of the Subcommission on Quaternary Stratigraphy (SQS), in short 'Anthropocene Working Group' (AWG), with the mandate to

---

This chapter is authored by Martin Bohle.

---

[1] http://www.igbp.net (consulted 12 May 2024)

[2] In the following, the terms Geosciences and Earth Sciences are synonyms. Geology and Earth System Sciences are disciplines within Geosciences.

M. Bohle et al., *The Anthropocene Working Group and the Global Debate Around a New Geological Epoch*, Emerging Globalities and Civilizational Perspectives, https://doi.org/10.1007/978-3-031-85175-9_2

> "examine the potential value of the term [Anthropocene] and its underlying stratigraphic basis in more detail and to consider, and subsequently make recommendations on, its possible formalisation" [ICS Annual Report 2010, p. 7].[3]

Hence, the AWG is a scientific body of an international non-governmental organisation (INGO) of geological stratigraphers. Its mandate is to study whether a geological formalisation of the concept of an Anthropocene is possible (Zalasiewicz et al. 2023). According to its self-assessment, the AWG undertakes to "critically consider the case for a formal Anthropocene, and to make recommendations to our parent [bodies]" [AWG Newsletter 2009, p. 1].

Given the ICS's statutory rules for working groups, the AWG ceased to exist on 31st October 2023, when it submitted its formal proposal (Waters et al. 2024a, 2024b) to the SQS.[4] Some months later, the SQS rejected the AWG's proposal in March 2024 (Witze 2024a). Subsequently, the ICS and IUGS management endorsed this decision (IUGS 2024; Witze 2024b).

At least temporarily, the ICS/IUGS decision halts the formal intra-organisational debates regarding modifying the GTS. The IUGS scheduled discussions on related geo-societal questions at the 37th International Geological Congress (August 2024 in Korea). The discussion of the geological and cultural implications of conceiving an Anthropocene continues, as publications made in the wake of the IUGS decision indicate (Koster et al. 2024; Zalasiewicz 2024; De Jouvancourt 2024; Ellis 2024; Barnosky and Hannibal 2024).

This chapter describes the dealings of the AWG as a working group in the ICS/IUGS system, including events that led to its creation in 2009.[5] What followed the AWG's proposal submission is mentioned but discussed in Chap. 6. For the present study, the newsletters of the AWG and the public reports of geological INGOs were inspected for features such as economic, institutional, interpersonal and intellectual resources, cooperation events, prospects to engage and positioning of individuals or scientific bodies. Scientific-technical questions (e.g. geological methodologies) and cultural implications of the notion of an Anthropocene are addressed if needed to understand the AWG's functioning but are not investigated further.

Scientific publications are ancillary sources when they are vehicles to express protagonists' positions. Occasionally, the absence of information is interpreted as

[3] For reference in square brackets, see Annexs in the Back Matter of the Book. Two tables list information sources (reports, newsletter) accessible online. All documents have been downloaded from the respective site. IT security software has identified several sites as not secure. If not labelled differently, these sources were consulted last on 31 January 2023 or 21 June 2023. Some sites were reorganised in 2024 following the ICS/IUGS decision to reject the AWG's proposal, and re-established later (accessible through: https://quaternary.stratigraphy.org/working-groups/anthropocene); archives exist at third locations.

[4] „A formal proposal was submitted to the SQS on October 31, 2023 that the Anthropocene be accepted as a stage/epoch terminating the Holocene, with a GSSP in a core recovered from Crawford Lake, near Milton, Ontario, Canada that would also define the Crawfordian Stage/Age." https://quaternary.stratigraphy.org/working-groups (accessed 16. February 2024; site reorganised later)

[5] I want to thank my co-authors of this volume, Boris Holzer, Leslie Sklair and Fabienne Will, for their valuable input and suggestions.

intentional. Also, hints to the emotional features of the Anthropocene debates are noted, e.g. stratigraphers from North America were wondering whether the

> "subjects of her Majesty are opinionated people'[6] and observe 'since then, have multiplied their articles to this effect. Above all, they inundate the media with messages, having well understood that communication prevails over reason and/or reflection…." (De Wever and Finney 2018, p. 255).

Likewise, understatements are identified, e.g. when the AWG chair describes the troubled year 2016 as

> "something of a transition interval for the AWG - negotiated, we hope, reasonably successfully. The end of one era for the AWG—if one may put it that way—came with the International Geological Congress at Cape Town in August 2016, where Colin Waters presented our main preliminary findings and interim recommendations (the adjectives are necessary in both instances, given that this is work in progress)" [AWG Newsletter 2016-17, p. 3].

Phrasings (as quoted) are intentional, indicate what might merit attention and illustrate nicely the post-normal conditions (Funtowicz and Ravetz 2003) of Anthropocene science.

2.2 Material and Methods

Earth Sciences or Geosciences study the Earth, mainly abiotic features, without neglecting biological and social attributes. The study subject (Earth) determines the characteristics of the specific knowledge-building practices of Earth Sciences, including those particular to Geology (Frodeman 1995; Seddon 1996; Baker 1999; Phillips 2012).

It is beyond this chapter's scope to situate knowledge-building practices in Geology within the broader set of Earth Sciences practices, including disciplines with profound physical and mathematical foundations, such as Meteorology. However, it is worth mentioning that the AWG uses knowledge from Geology and Earth System Sciences (ESSs). Both disciplines are at opposite ends of a spectrum of the mathematisation of Earth Sciences.

The recent debate on whether the Anthropocene should be formalised as an epoch or event (Bauer et al. 2021; Gibbard et al. 2021; Waters et al. 2022; Edgeworth et al. 2023) illustrates the societal relevance of Earth Sciences, including Geology (Hall 1976; Zen 1993; Schofer 2003; Ismail-Zadeh 2016; Warde et al. 2017).

The notion of Anthropocene caused ample public interest (Brauch 2021; Sklair 2021), as described with emphasis on the AWG in Chap. 5. Furthermore, the AWG's findings have implications for many scientific communities (Trischler 2016; Lundershausen 2018b; Barton 2019; Wenninger et al. 2019; Wallenhorst 2023), as illustrated in Chap. 4, and despite the AWG's limited 'stratigraphic mandate'. Gauged by the latter, the intellectual outreach of the AWG was very substantial, as illustrated in Chap. 3. The dealings of the AWG, as described in the present chapter,

[6]The wording 'subjects of her Majesty' refers to the leading function of British scientists in the AWG.

follow a sequence of inclusions and exclusions (Lundershausen 2018c) along a specific path seeking peer consensus, which has been found only with some peer groups (Chap. 6), e.g. excluding the geological mainstream as represented by the IUGS.

2.2.1 Analytical Lenses

The two concepts of a 'synthetic thesis of truth' (Mario A. Bunge; 1919–2020) and an 'evolution of knowledge' (Jürgen Renn; 1956-) guide the following analyses.

These two concepts fit the knowledge-building practices of Earth Sciences because experiments are possible only for some phenomena, and hypothesis testing is constrained by the limited reproducibility of natural/social phenomena. Instead, repeated observations of recurrent phenomena (e.g. volcanic eruptions) and mathematical simulations (e.g. weather, climate) support scientific insights. Given these circumstances, scientists apply the practice of a synthetic thesis of truth[7] (Marone et al. 2019), i.e. synthesising substantial amounts of practical knowledge with chunks of theoretical knowledge (sometimes in mathematical form). The concept of an 'evolution of knowledge' describes the underpinning social practices.

The Earth Sciences strive for discipline-specific standardisation, including mathematical formalisation if possible. The GTS is an example of standardisation in Geology (Gradstein and Ogg 2020), and the ICS has a specific mandate to maintain its integrity [e.g. ICS Annual Report 2022, p. 1]. Standardisation in ESS is driven by mathematisation, data science and numerical modelling. The article, 'The trajectory of the Anthropocene: The Great Acceleration' (Steffen et al. 2015) provides an example.

Geological Stratigraphy and ESS represent opposite configurations (Baker 1999) of the knowledge-building practice of Earth Sciences, one being not and the other much mathematised. Therefore, the AWG is challenged (Steffen et al. 2016) to combine both configurations to derive a geological Anthropocene concept that is sound in terms of stratigraphical methods:

> "The AWG may initially have welcomed the concerns of various disciplines, and invited proposals from them, but this was ultimately toward the end of adopting one position that was as comprehensive as possible, at the exclusion of all other possible positions. The position adopted by the AWG (namely that the Anthropocene is a legitimate unit at the rank of Epoch/Stage, which should have a lower boundary defined by a GSSP using the mid-twentieth century signals associated with the Great Acceleration as a primary guide) is not by any account the best of all possible options, but simply that which succeeded in accordance with disciplinary protocol the AWG are subject to as a constituent body of the ICS and SQS." (Damianos 2021, p. 125, 126)

---

[7] A Synthetic thesis of truth "requires considering a hypothesis corroborated both by purely empirical confirmation and external consistency or compatibility with the bulk of existing background knowledge (systemicity)…. Pattern consistency (empirical control) together with an understanding of causal relations (rational together with empirical control) make confirmed hypotheses robust and more reliable" (Marone et al. 2019, p. 363).

Concepts taken from Renn's theory of the evolution of knowledge (Renn 2020), such as the 'economy of knowledge', illustrate the knowledge-building practices of the AWG. An 'economy of knowledge' is the set of societal processes and material circumstances, i.e. external representation, which determines how knowledge changes in institutions, societal practices, and cultures.[8] The AWG, including its institutional and cultural embedding, belongs to an economy of knowledge for the sciences of planetary-scale anthropogenic change ('Anthropocene knowledge'). The AWG is part of the material/external representation featuring this specific economy of knowledge. Likewise, the scientific-bureaucratic embedding of the AWG in the ICS/IUGS structure belongs to this material/external representation. Within an economy of knowledge, borderline problems[9] drive the evolution of knowledge (Renn 2020, p. 81). Borderline problems describe concepts (or objects) that possess features belonging to different knowledge domains, which cause the integration and reorganisation of the original knowledge.[10] Using Renn's theory, the AWG can be understood as an 'epistemic island',[11] i.e. an identifiable community within broader knowledge systems that tackles specific borderline problems.[12]

Multiple borderline problems are embedded in the concept of a geological Anthropocene, for example, the 'onset' of the epoch, i.e. the stratigraphically discernible features related to human activities such as the Orbis spike (Lewis and Maslin 2015). The Orbis spike is a brief drop in the atmospheric carbon dioxide concentration, which is related to the reforestation in North America after the eradication of endogen American cultures (Denevan 2011). Likewise, the testing of nuclear weapons in the atmosphere led to a distinct pattern in the stratigraphic record. Other borderline problems of the Anthropocene concept are, for example, the role of conceptual features such as system shifts or planetary boundaries. Other borderline problems relate Earth Sciences to Social Sciences, such as the question of human agency in historical and socio-economic contexts (Will 2021; Thomas 2022; Wallenhorst 2023; Ellis 2024). Likewise, the geological event concept, the geological episode concept and the geological epoch concept (Bauer et al. 2021; Gibbard et al. 2021; Waters et al. 2022) describe specific borderline problems in Anthropocene science. Each concept aggregates selected knowledge, prioritising

[8] "all societal processes pertaining to the production, preservation, accumulation, circulation, and appropriation of knowledge mediated by its external representation" (Renn 2020), p. 429), whereas external representation means "any aspect of the material culture or environment of a society that may serve as an encoding of knowledge", (Renn 2020, p. 224)

[9] Borderline problems are (Renn 2020, p. 427): 'Challenging objects or problems that belong to multiple distinct systems of knowledge. Borderline problems put these systems into contact (and sometimes into direct conflict) with each other, potentially triggering their integration and reorganisation'.

[10] The rise of modern Earth Sciences (Geosciences) is driven by borderline problems (Renn 2020, p. 237–240).

[11] See (Renn 2020) for a glossary of notions like epistemic island, knowledge system, borderline problems.

[12] understood to be a suite of challenging subjects belonging to multiple distinct systems of knowledge.

different cultural, historical, socio-economic and geological features and 'user' communities (Bohle 2023). The AWG's choice is to combine into a specific 'borderline problem' a geological methodology (the Geological Time Scale) with the understanding that the Earth System dynamics shifted (Turner et al. 2024). The IUGS's choice, formalised in the decisions taken in 2024 (IUGS 2024), is different.

2.2.2 *Triangulations*

In the following, the dealings of the AWG will be triangulated from three viewpoints.

> First: The AWG is a body within an autonomous international scientific organisation (an established INGO) created by geologists to coordinate and represent their disciplines. The Subcommission on Quaternary Stratigraphy (SQS) of the International Commission on Stratigraphy[13] (ICS), which set up the AWG in 2009 to tackle a specific question of geological terminology/methodology, is a third-layer sub-body within the International Union of Geological Sciences (IUGS). The IUGS, in turn, is a member of the International Science Council (ISC), the world's leading non-governmental organisation of international scientific unions and associations, including different organisations representing the Earth Sciences. Hence, the AWG is a minor part of an established non-governmental scientific bureaucracy with a specific disciplinary focus.
>
> Second: The IUGS established the Geological Time Scale (GTS) and the International Chronostratigraphic Chart (ICC) in the second half of the last century as scientific resources comprehensively described by geological methodologies. The IUGS supports this achievement, takes epistemic leadership and oversees any related issue. The AWG was mandated to examine whether the GTS should be amended. The amendment would imply ending the geological epoch of the Holocene in the historical past and adding a new geological epoch at the end of the GTS, the Anthropocene. Agreeing on the Holocene, including its internal structure, was a significant endeavour of geologists for most of the past 200 years. Hence, proposing the Anthropocene disrupts the established GTS.
>
> Third: Earth System Scientists proposed the change to the geological terminology in 2000. It provoked multiple debates in various scholarly communities and was done with little concern for the methodological issues of Geology. Furthermore, the term Anthropocene seemed erroneous to many humanities and social science scholars because it refers to a non-differentiated and non-historical human agent (*Anthropos*). Finally, asserting that human actions modified Earth's geological features is not an intrinsic geological question. Hence, the Anthropocene is a massive borderline problem, and the proposal is an incident of epistemic intruding.

Combining these perspectives helps to understand how the AWG developed, including the interaction with the cultural/public sphere, for example,

---

[13] https://www.iugs.org/ics 'The International Commission on Stratigraphy is the largest and oldest constituent scientific body in the International Union of Geological Sciences (IUGS). Its primary objective is to precisely define global units (systems, series, and stages) of the International Chronostratigraphic Chart that, in turn, are the basis for the units (periods, epochs, and age) of the International Geologic Time Scale; thus setting global standards for the fundamental scale for expressing the history of the Earth'.

"[t]here is the continuation of the major exhibition at Berlin's Haus der Kulturen der Welt, the preparations for the equally significant Deutsches Museum/Rachel Carson Anthropocene exhibition, and a little farther into the future, the incorporation of the Anthropocene as a major theme in the Smithsonian Institution's forthcoming 'Deep Time' exhibition." [AWG Newsletter 2013-14, p. 1].

2.2.3 *Information Sources*

INGOs such as the ICS or the IUGS publish annual activity reports. The AWG did publish regular newsletters, which should be distinguished from peer-reviewed scientific articles or books by AWG members and internal reporting.[14] These reports and newsletters are issued under the responsibility of leading individuals (officers) as required by statutes, by-laws or practices. In addition to this regular information, some additional information can be found, for example, in blogs like the 'SQS news[15]', addressing specific audiences under the respective authors' responsibility. The AWG's newsletters may include sections that are formal reports to the SQS management; at least, the chairperson's statement reads like it. There was no evidence that the AWG made regular specific reports to the SQS.

The regular activity reports (including newsletters) for 2009 to 2023, published to describe the dealings of the respective geological INGOs, inform this study. Statutes and by-laws of the scientific bodies or organisations contextualise this reporting. Publications of leading individuals (officers) give ancillary information because they frequently outline appreciation (of events and intentions) in the margin of scientific discussions. Also, scientific organisations organise publications governed by editorial boards supporting preferred perspectives, including choices on which scientific viewpoints merit visibility. Hence, the publication profiles of journals provide some additional evidence.

Most functions (offices) in INGOs are assumed voluntarily, and the office has limited material resources or funding, e.g. [ICS Annual Report 2022, p. 4]. This feature indicates how to read activity reports and newsletters. Esteem, commitment and interpersonal relationships are essential to fulfilling offices.[16] For some individuals, working for INGOs has been part of their professional identity for many years. Subsequently, these individuals shape the organisation, which, however, is a

[14] The ICS's rules for reporting are 'Chairs of Task Groups and Ad Hoc Committees shall transmit annual reports to the chair of the ICS body under which they reside. They are also responsible for including the group's operating costs in that report and projected budget for the new reporting period'. [§13.1 second Paragraph].

[15] http://quaternary.stratigraphy.org/category/sqs-news/ (accessed 16. February 2024; site reorganised later)

[16] For example, [ICS Annual Report 2018, §10]: 'The work of SQS is financed almost exclusively through research grants of individuals working on prospective GSSP sections. Such funding is obtained primarily from national and regional governmental research agencies, institutes, and universities. The Anthropocene Working Group has been particularly successful in obtaining funding owing to the high visibility of the Anthropocene topic'.

function attributed to them by their peers. Therefore, the official reports and newsletters are also personal records of what officers consider relevant for their peers. Hence, beyond having a public and institutional nature, the reports and newsletters are a vehicle to communicate individual commitment. Therefore, they permit analysing events and processes if understood as a biased means for purposeful reporting to specific audiences.

Given that some individuals were outspoken, some incidents mentioned in the following sections may be perceived as interpersonal quarrels over value statements. Even if this tilt should be accurate, it is part of the studied properties of the professional environment. The specific subject of the AWG had taken the form of a geological (scientific) question. However, the broader (borderline) problem of an Anthropocene to which the geological question belongs involves value statements and political preferences (Finney and Edwards 2016; Koster 2022; Koster et al. 2023; Koster et al. 2024). Hence, interpersonal quarrels about value statements do not seem extraordinary.

Under the aforementioned caveats, events relevant to the AWG's dealings are listed and analysed in the following sections, such as circumstances encountered, seized opportunities and actions taken. The main event timeline covers 2008–2023, including some events to understand AWG's emergence and the handling of AWG's proposal by SQS, ICS and IUGS. The analysis is limited to publicly available reporting. Other approaches, for example, interviewing AWG members (Will 2021) or embedding (Damianos 2021), have been applied elsewhere and inform Chap. 4 of this book.

2.3 The AWG's Environment

The AWG is not a scientific body *sui generis*. Instead, it is part of an INGO, internationally representing communities of geologists (Dodick and Orion 2003) (Cohen et al. 2013; Finney 2014; Wenninger et al. 2019 [p. 35–43]; Foster 2018; Lundershausen 2018a; Damianos 2021; pp. 90–108). The following subsections outline four perspectives featuring the AWG and illustrate 'entanglements'.

The AWG's emergence and functioning follow well-established practices of the international geological science communities (Schofer 2003). In this environment, interpersonal networks and voluntary contributions are essential for operability. For routine business, the intra-organisational hierarchy seems better described by terms like constituent rather than subordinated or consent rather than guidance. Interfaces with the global networks of the intergovernmental UN/UNESCO science system are multiple. Leading and often long-term-serving individuals navigate this multi-layer environment guided by the practices of their communities.

2.3.1 *The Dawn of the AWG*

In 2017, Jan Zalasiewicz and some of his peers published an online contribution entitled 'The Geological Anthropocene born in Burlington House' that illustrated the dawn of the AWG:

> "…to Burlington House, 17th May 2006, and a meeting of the Geological Society of London's Stratigraphy Commission [[17]]. One of us (JZ), as its (then) chair, introduced a draft manuscript on the Anthropocene for consideration as a Commission publication. The term was just beginning to get enough visibility for stratigraphers to take notice, and its use in literature emerging from the [International Global Geosphere-Biosphere Program] IGBP was generally matter-of-fact—as if it were already an accepted term. However, it was still emphatically informal,… The Stratigraphy Commission [of the Geological Society of London] is not an international body, so cannot "adjudicate"—though it can pass comment. Perhaps somewhat to its own surprise, it agreed (by large majority) that the Anthropocene seemed "geologically reasonable" and might be considered for formalisation.
>
> This Commission's cautiously-worded paper on the Anthropocene (authored by all but one of the 22 Commission members) was published in 2008 by the Geological Society of America (Zalasiewicz et al. 2008). It attracted attention…. As another result, there came an invitation from one of us (PG, then Chair of the Quaternary Stratigraphy Subcommission of the International Commission on Stratigraphy (ICS)), to set up a body (the Working Group on the "Anthropocene", to which all five authors belong) to formally examine the evidence as to whether the Anthropocene might become part of the Geological Time Scale, and to make recommendations." (Zalasiewicz et al. 2017c).

In a nutshell (and published 10 years after the event), this quote illustrates the environment in which the AWG was set up. The science behind the Burlington House event likely stems from the activities of British stratigraphers to revise the description of Quaternary deposits in the British Isles (Bowen 1999) and to determine the general status of the Quaternary in the geological nomenclature (Gibbard et al. 2005). The stratigraphers understood better the most recent features of the regional Quaternary deposits (Price et al. 2011) and how they relate to insights of the IGBP.

Aside from the GSA paper (Zalasiewicz et al. 2008) mentioned in the quote, Jan Zalasiewicz published the book 'The Earth After Us—What legacy will humans leave in the rocks?' (Zalasiewicz 2008). He develops for an interested public a science-fiction narrative of geological Stratigraphy of the human age in 100 million years, briefly addressing the Anthropocene question. He concluded 'a reasonable interim aim might be to try very, very hard for the Anthropocene to develop into something no greater than an epoch-scale event' (Zalasiewicz 2008, pp. 155–157). This publication strategy likely foreshadows the approach of the AWG to coupling engagement in the scientific and public spheres. Likewise, the little quote aforementioned may say something about self-motivation.

[17] https://www.geolsoc.org.uk/Groups-and-Networks/Commissions/Stratigraphy-Commission

2.3.2 INGOs of the Geological Communities

The international, non-intergovernmental geological community is organised into a hierarchy of enmeshed INGOs. Damianos describes them from the perspective of the AWG (Damianos 2021; pp. 89ff.). Lundershausen and Will describe the underlying cognitive processes (Lundershausen 2018a; Will 2021).

The two top-level INGOs are the International Union of Geological Sciences[18] (IUGS, founded in 1961) and the International Union for Quaternary Research[19] (INQUA, founded in 1928); the former is a heavyweight in number of individual geologists to whom it can reach out.[20] The IUGS is the main sponsor of the International Geological Congress (IGC), first held in 1878 (Ellenberger, 1978), occurring every 4 years. This historical root assigns IUGS with seniority, although it is a latecomer among scientific INGOs. The IUGS coordinates national geological organisations, often represented by intermediates (e.g. a national committee of the IUGS). The IUGS successfully participates in intergovernmental cooperation, such as the International Geoscience Programme[21][22] of the UN/UNESCO agencies, such as the World Meteorological Organisation (WMO).[23]

IUGS and INQUA are members of the International Science Council (ISC), which organises a global spectrum of disciplines and scientific organisations alongside the intergovernmental UN/UNESCO system. From an Earth Sciences perspective, the range of disciplines and scientific organisations of the ISC is completed by the International Union of Geodesy and Geophysics (IUGG), which represents geophysical, meteorological and oceanographic sciences (Ismail-Zadeh 2016). Although the sciences represented by IUGS, INQUA and IUGG belong to the scope of ESS, inter-organisational cooperation seems limited.

The IUGS and the INQUA have overlapping membership and disjunct interests. The INQUA is oriented towards environmental sciences and holds a broad spectrum of disciplines, including Geology and social sciences.[24] The IUGS considers geology to be the centrepiece of its concerns.[25] The INQUA focuses on the recent ~2.6 million of Earth's geological past, the Quaternary (Gibbard and Head 2020). The IUGS focuses on the entire geological history of Earth (4.6 billion years).

Recently, relations between the IUGS and the INQUA have shown tensions about geological terminology that describes the Quaternary. Two decades ago,

---

[18] https://council.science/member/iugs-international-union-of-geological-sciences/

[19] https://council.science/member/inqua-international-union-for-quaternary-research/

[20] See also International Union of Geological Sciences. 2012. IUGS Book of Facts, 1961–2011; https://www.researchgate.net/publication/280041640_IUGS_Book_of_Facts_1961_-_2011

[21] https://www.iugs.org/igcp

[22] https://en.unesco.org/international-geoscience-and-geoparks-programme

[23] https://public.wmo.int/en

[24] https://www.inqua.org; https://www.inqua.org/media/uploads/5cdd78d8aafff_INQUA_mission_statement.pdf

[25] https://www.iugs.org/history

geological stratigraphers organised in the ICS intended to replace the term Quaternary in the official geological terminology (Gibbard et al. 2005; Will 2021, pp. 73–84). The IUGS and the INQUA found a compromise in 2009, maintaining the Quaternary (composed of the Pleistocene and the Holocene) within the GTS. Preparing for this compromise, the ICS cooperated with INQUA to re-establish the SQS in 2002. Jan Zalasiewicz and Mark Williams, in a book about Earth's climate history for the interested public, published in 2012, summarise the 'traumatic' battle

> "over a definition of a geological term: The Quaternary… Part of the geological community… not least to preserve the stability of the Geological Time Scale. But, most scientists working on Ice Age phenomena were unhappy… There followed the most almighty kerfuffle… Many of those involved (we too) still bear the scars, though this definition[26] seems now set to stay." (Zalasiewicz and Williams 2012; p. 160).

Since the 1970s, the ICS and IUGS have established the GTS (Cohen et al. 2013; Gradstein and Ogg 2020). This primary accomplishment provides a valuable resource for geology and other sciences. Within this endeavour, the stratigraphic features of the Holocene were settled as late as 2018.[27] The proposal of a geological epoch Anthropocene (Crutzen and Stoermer 2000; Crutzen 2002) made outside the IUGS structure and the subsequent establishment of the AWG as a constituent part of the IUGS structure (Zalasiewicz et al. 2017b) overlay the process of agreeing on the stratigraphic description of the Holocene.

2.3.3 *Establishing the AWG*

The SQS shall apply the statutory scientific approach of the ICS to the stratigraphy of the Quaternary.[28] Consequently, the SQS established a working group on the Anthropocene in 2009. It works in parallel with the SQS's working groups about the geological subdivision of the Holocene. Also, the INQUA has a commission dedicated to the stratigraphy of the Quaternary. Beyond these international commissions, various national and institutional bodies exist, such as the Stratigraphy Commission of the Geological Society of London. The interaction of these multiple bodies is not transparent to outsiders, although the IUGS/ICS has leadership for issues of geological terminology. Finally, the role of the IUGG is unclear in this context, although it represents an essential part of the user community of geological terminology.

The ICS statute frames the AWG's working modes. The statute specifies duration, membership, reporting and voting rules. Generally, the mandate of a working group is 'the selection and definition of the lower boundaries of chronostratigraphical/geochronological unit' [§7 of ICS statutes].

---

[26] Quaternary Period began a little less than 2.6 million years ago.

[27] http://quaternary.stratigraphy.org/iugs-ratifies-holocene-subdivision/

[28] http://quaternary.stratigraphy.org (accessed 16. February 2024; site reorganised later)

The lifetime of a working group is 4 years and renewable once. Optionally, a working/task group is reconvened after 8 years by the chair of the ICS subcommission. The responsible ICS subcommission selects (non-paid) officers (chair and secretary) of a working/task group. The working/task groups agree on their members; voting members need confirmation from the ICS management. The function of working/task groups or members who do not communicate may be terminated. Decision-taking (voting) is by 60% majority and 60% quorum. The processes for re-constituting the working/task group after 4 and 8 years and nominating the voting members are the main levers of the ICS to shape its working groups.

The ICS describes the process of setting up the AWG in the 2009 report:

> "The Anthropocene Working Group, chaired by Dr J. Zalasiewicz (Leicester) was created in the summer of 2009, following the proposal of the term Anthropocene by (Crutzen 2002), its subsequent analysis by the Stratigraphy Commission of the Geological Society of London (Zalasiewicz et al. 2008) and a session convened at the 2008 Fall Meeting of the American Geophysical Union on this theme. The purpose of the Working Group is to examine the term and its underlying stratigraphic basis in more detail and to consider, and subsequently make recommendations on, its possible formalisation." [ICS SubComm Report 2009, pp. 4–5].

Early in its existence, the AWG did not consider procedural aspects and the IUGS/ICS context. In 2014, the AWG discussed them [AWG Newsletter 2014, p. 3], introducing the "[k]ey questions identified by the International Commission on Stratigraphy:

- Is there a well-documented and significant stratigraphic record for the "Anthropocene" on which to base its recognition worldwide and to define a boundary with underlying Holocene strata?
- Should the base of the "Anthropocene" be defined using a physical reference section ("golden spike" or GSSP) or instead should its beginning be defined in terms of a numerical date (GSSA)?
- Is the "Anthropocene" a unit of Earth history or human history?
- What is the usefulness of the "Anthropocene" as a time unit and a material unit to be visualized by geoscientists and other interested communities?
- What is the relative value of formalizing the "Anthropocene" chronostratigraphic/geochronological unit as opposed to leaving it as an informal term?"

The AWG reports this detailed task description, not the ICS. It prescribes issues traceable in scientific publications throughout the AWG's lifetime.

2.3.4 *Resourcing Geological INGOs*

Typically, the geological INGOs are cash-poor. For example, annual funding 2021 for the IUGS (~527.000 US$) comes mainly from IUGS membership fees (~75%) and a grant from UNESCO (~16%) [IUGS 2021 report, p. 9]. Regarding 2020 expenses, funding of IUGS commissions and the ICS are the two most essential items (72.200 US$, 47.000 US$) out of a total budget of 332.610 US$ [IUGS 2021 report, p. 10]. ICS activities are funded by an annual grant[29] from the IUGS [ICS Annual Report 2021]. Access to support 'in kind' is crucial for any ICS working group, the ICS subcommissions and the ICS's operability. The ICS subcommissions and working groups must seek external funding for their tasks, and members might use the resources from their professional affiliations (beyond dedicated time).[30] Hence, the professional sphere and ICS activities are tied, and only those activities flourish that officers or working group members can justify within their respective professional duties. From a functional perspective, this frugal environment gives the geological community options for checks and balances to keep the INGOs operating within margins.

The AWG managed to mobilise the resources essential for its functioning, although obtaining dedicated financial funding proved difficult (Damianos 2021; Annex 1). Applications for institutional financial support through established science funding structures seemed unsuccessful[31] (Damianos 2021, p. 276); it might have also been deterred by worries about taking a supposed (official) position. Instead, organisations and institutes inside and outside the geological community seemed interested in providing 'in kind' assistance to the AWG, mainly by hosting meetings. Also, AWG publications have been funded through 'in kind' support. The total amount of resources bundled around the AWG's activities is unknown. Substantial financial support for GSSP fieldwork organised by *Haus der Kulturen der Welt* (HKW) in Berlin (Germany) came in later years.[32] It had specific boundary conditions, which were ingeniously used to combine scientific study and public visibility (Damianos 2021; [p. 229ff], Koster 2022).

2.4 The AWG's Functioning: Building Blocks

Given the role of individuals within geological INGOs, the following five subsections address the building blocks for describing the AWG's critical features and then its development stages. The building blocks are leadership, membership, external

[29] 'The IUGS Executive Committee awarded ICS a budget of $40,000 for 2020, reducing significantly the requests of the subcommissions and the ICS Executive. No additional income to the ICS is declared'. *§9;* the amount differs from the amount given in the IUGS report.

[30] 'Informally, every officer and member of ICS donates their own time, office space, institutional facilities, and other components to the activities of the organisation. No officer nor executive receives any salary compensation from IUGS or other ICS funds. Indeed, most officers personally contribute towards their own travel and operational expenses'. [ICS Annual Report 2022, p. 4]

[31] The AWG Newsletter 2011 [p. 2] mentions an application to NERC for supporting an international network. No reference in later newsletters exists on the fate of this application.

[32] 'The Anthropocene Working Group has been particularly successful in obtaining funding owing to the high visibility of the Anthropocene topic. However, it is difficult to obtain funding for SQS conference travel and working group field trips, and ICS funding is accordingly critical to the activities of SQS'. [ICS Annual Report 2019, p. 18]

players, IUGS' attitudes and meetings and publications) as the primary means of communication for the AWG.

2.4.1 Leadership

The IUGS and ICS statutes[33][34] give their voluntary officers[35] time-limited mandates within a system of complex representations, checks and balances, including (lasting) inter-organisational quarrels like conflict about the stratigraphy of the Quaternary[36] between IUGS and INQUA.

This conflict ranged from whether to use the term 'Quaternary' over what the Quaternary's stratigraphic division is to determine the subdivision of the Holocene. The latter issue impinges on the Anthropocene question. The available reporting gives limited insights into the current cooperation between the IUGS/ICS/SQS and the INQUA. However, their collaboration is formalised at the level of the ISC,[37] practised[38] and secured by cross-membership.[39]

To illustrate the direct involvement of individuals, scientists shaping the environment of the AWG (e.g. Stan Finney or Phil Gibbard) have been officers active in settling the compromise about the Quaternary at the 33rd IGC in Oslo 2008 and, subsequently, agreeing on the subdivision of the Holocene in 2018.[40] A decision on an Anthropocene epoch would imply that the Holocene ends, which is a contentious proposal for some (Gibbard and Walker 2014; p. 35) and, hence, relates the Anthropocene question to the former quarrel about the fate of the Quaternary. The hands-on handling of the IUGS/ICS management to obtain in March 2024 a vote from SWQ on the AWG proposal demonstrates the influence of individuals [AWG Newsletter 2023; p. 4] (Witze 2024a).

[33] https://www.iugs.org/statutes-bylaws

[34] https://stratigraphy.org/statutes (accessed 16. February 2024; site reorganised later)

[35] often senior scientists with status of Emeritus

[36] http://quaternary.stratigraphy.org/definitions/correspondence/ (accessed 16. February 2024; site reorganised later)

[37] INQUA member since 2005, and IUGS since 1961

[38] http://quaternary.stratigraphy.org/definitions/correspondence/ (accessed 16. February 2024; site reorganised later)

[39] 'The Subcommission on Quaternary Stratigraphy (SQS) maintains close ties with The International Union for Quaternary Science (INQUA) through its Commission on Stratigraphy and Chronology (SACCOM). SACCOM supports research on candidate GSSPs through its grants program, so directly assists the work of SQS. Of equal importance, INQUA established, during its congress this summer, a permanent committee with the express task of liaising with SQS. This will allow the INQUA Executive to respond rapidly to any issues of stratigraphy raised by SQS and vice versa. Prof. Thijs van Kolfschoten, a Vice-President of INQUA, chairs this committee, and also serves as a voting member of SQS'. [ICS Subcommission Annual Reports for 2015, §9C] (https://stratigraphy.org/files/ICS_SubcommReport2015.pdf)

[40] https://stratigraphy.org/news/121

The number of officers visibly involved in the dealings of the AWG is tiny:

- Jan Zalasiewicz assumed the function of the chairperson of the AWG from 2009 to 2019, followed by Colin Waters. Colin Waters was AWG secretary from 2011 to 2019, preceded by Mark Williams and followed by Simon Turner. The latter is the coordinator of the GSSP[41] project of the HKW, supporting the work of the AWG from 2019 onward.
- Stan Finney chaired the ICS during the first years of the AWG up to 2016, followed by David Harper. After leading the ICS, Stan Finney was secretary general of the IUGS from 2016 onward. Philippe Gibbard, a member of the AWG since 2009, was chair of the SQS until 2011, followed by Martin Head. Since 2016, Phil Gibbard has been the general secretary of the ICS.
- The long-serving chairperson of the AWG, Jan Zalasiewicz, blended this function with positions in the ICS. From 2009 to 2015, he was vice-chair of the Subcommission in Stratigraphic Classification (SSC), and from 2012 to 2019, he was secretary of the SQS and later its chair, succeeding the 2012–2020 chair Martin Head. Martin Head joined the AWG probably after the 35th IGC, although he cooperated with the AWG before the conference.[42] In 2020, Martin Head was one of two vice-chairs of the SQS.

The replacement of individuals in the pecking order of the AWG, SQS, ICS and IUGS is probably driven by a combination of statutory requirements, the limited number of peers having interest and standing to assume officer functions in INGOs and opportunities to access higher organisational levels of decision-making for matters considered vital, i.e. position taking for power plays. Evidence for this assumption is found in Jan Zalasiewicz's move to take the chair of the SQS when this body would receive the AWG proposal. Likewise, it seems more than likely that the scientific views of Stan Finney[43] and Phil Gibbard[44] regarding the appropriateness of a geological epoch Anthropocene (see respective publication records) played a role in the IUGS's management choices (Witze 2024a) overruling the SQS Chair (J. Zalasiewicz).

2.4.2 Membership

Throughout its existence, the membership of the AWG was mainly male and geographically skewed, although disciplinary diversity is noticeable (Brauch 2021, p. 357f) as also detailed in Chap. 3.

[41] Global Stratotype Section and Point (GSSP)

[42] Attending third meeting of the AWG (22–23 April 2016)

[43] Post on Facebook (fifth March 2024, 13, 477) "It's done. The Anthropocene epoch rejected"

[44] Left the AWG in 2023.

At its inception in 2009, the AWG comprised 16 people, including Paul Crutzen,[45] the *spiritus rector* of the idea to end the Holocene and introduce a new geological epoch, the Anthropocene. At that time, each second AWG member was affiliated with institutions in the United Kingdom, and all members except one were male. Of those individuals who started the work of the AWG, eight are still members of the AWG in 2019, including three voting members (Phil Gibbard, Mark Williams and Jan Zalasiewicz). Other five AWG members worked from 2009 onwards, including Erle Ellis[46] and Will Steffen.[47] Both have distinct (non-Geology) scientific profiles. For example, the environmental scientist Erle Ellis is known for his ecomodernist views (Asafu-Adjaye et al. 2015) and specific opinions (Bauer et al. 2021). The IGBP scientist Will Steffen (Hughes and Rice 2023) is known for his work on planetary boundaries and the Great Acceleration (Steffen et al. 2015). Eric Odada, professor of Quaternary Geology at the University of Nairobi, is one of the two African scientists who participated in the AWG between 2009 and 2022. An Zhisheng, an AWG member since 2009, is affiliated with the Institute of the Earth Environment of the Chinese Academy of Sciences. He is one of the small group of scientists from Asia who participated in the AWG between 2009 and 2022. The fifth long-term serving member of the AWG is Davor Vidas, a law of the sea specialist from Norway who, in the summer of 2009, communicated with Jan Zalasiewicz about the formal status of the notion of an Anthropocene [AWG Newsletter 2009, p. 1].

In 2021, before the final round of formal voting, the AWG comprised 38 members, 23 voting and 15 advisory members. Eighteen of the voting members are male, and 17 of the voting members are affiliated with institutions located in European countries. Five voting members are affiliated with institutions in North America; two are affiliated with institutions in East Asia. Twelve of the advisory members are male, and six of the advisory members are affiliated with institutions located in European countries. Six advisory members are affiliated with institutions in North America, and one each from (east) Asia, Africa and Australia.

Taking language as an indicator of cultural diversity, about two-thirds of the AWG members come from English-speaking parts of the world. Likewise, the AWG's composition is skewed European-centric and towards white male participation. The skewness, which likely reflects the geological community's limited diversity, diminished over time, and it is particularly noticeable for the voting membership.

It is beyond this study's scope whether and how the AWG composition and dealings reflect the status quo of equity, diversity, gender and inclusion in the Earth Sciences communities (Mogk 2020; Baldwin and Erickson 2020), including a systemic unrepresented Global South.

---

[45] died 28 Jan. 2021

[46] Resigned in July 2023, https://anthroecology.org/why-i-resigned-from-the-anthropocene-working-group/ (accessed 23 July 2023)

[47] died 29 Jan. 2023

2.4.3 External Players

Beyond officers of the IUGS and the ICS, who affected the work of the AWG, some individuals shaped its development in a particular manner, be it as members of the AWG or as (external) promoters. The available reports, newsletters and statutes do not indicate sensibility for conflict-of-interest issues for individuals or organisations.

The late Paul Crutzen (AWG honorary member) has a particular role as a *spiritus rector* of the concept of an Anthropocene (Molina and Rowland 2010). Initially, he placed the notion of an Anthropocene publicly (Crutzen and Stoermer 2000; Crutzen 2002) and promoted it in publications and at meetings, including the Stockholm Memorandum (Laureates 2011; Folke and Rockström 2011). Likewise, his influence was felt after the 35th IGC as the AWG met at the Max Planck Institute for Chemistry, where Paul Crutzen was affiliated as *Emeritus*. The crucial role of Jan Zalasiewicz in the AWG is hard to underestimate, first as a kind of mid-wife to Paul Crutzen's idea (Zalasiewicz et al. 2017b), second as a scout of the complex landscapes of organised stratigraphy (Zalasiewicz et al. 2021) and third as a narrator of Earth's history (Zalasiewicz 2008; Zalasiewicz and Williams 2012) capable of bridging disciplines (Zalasiewicz 2022).

As outlined earlier in the 'leadership' subsection, some members of the AWG (Phil Gibbard, Martin Head, Jan Zalasiewicz) have also been officers of the ICS throughout the existence of the AWG. Being members of the AWG directly informs the ICS, and as officers of the ICS, they can directly shape the working environment of the AWG:

- Under Phil Gibbard's mandate as chair of the SQS, the AWG was installed.
- Under Martin Head's mandate as chair of the SQS, the AWG navigated the 35th IGC.
- Under Martin Head's mandate as chair of the SQS and Phil Gibbard's mandate as secretary general of the ICS, the AWG was re-established in 2020, with Colin Waters and Jan Zalasiewicz as incoming chairs of AWG and SQS, respectively.

Stan Finney, first the chair of the ICS and later Secretary General of the IUGS, is the leading protagonist of those parts of the geological community that is critically opposed to an Anthropocene as part of the GTS (see, for example, (Rull 2018; Edgeworth et al. 2019; Ruddiman 2019; Gibbard et al. 2021)). His views can be traced through statements/interventions in the context of the 34th IGC[48][49] and 35th

[48] 'Media interest in the "Anthropocene" kept the [ICSC] Chairman [Stan Finney] busy responding to interview requests'. [IUGS Annual Report 2012]

[49] 'Thus, since 2012, during the time of the Geological Congress held in Brisbane, we have seen journalists and scientists write that this point was going to be debated and voted on. Rumors came solely from the media, as this discussion was not on the agenda! The dossier was not scientifically informed. The president of the International Stratigraphic Commission (co-author of the present article) had specified that it would perhaps not even be ready before the next congress, in 2016. He was also surprised that this proposition was put forward by scientists who do not in fact all have a clear idea of what a geological subdivision represents and what one requires (in terms of landmarks, recording findings in sediments, etc.)' (De Wever 2018, p. 255)

IGC[50] and publications (Finney 2014; Finney and Edwards 2016; De Wever and Finney 2018). Finally, in 2021, Stan Finney, then Secretary General of the IUGS, joined the plea for an Anthropocene event instead of an Anthropocene epoch (Gibbard et al. 2021) published in the IUGS's main journal (Episodes). The role of S. Finney and Ph. Gibbard in the ICS/IUGS decision in March 2024 is non-documented but should not be underestimated. Formally, given the ICS/IUGS statutes, a conflict of interest is not given; considering the substance and context of the decision, the opposite could be stated.

Other influential individuals are affiliated with the INQUA,[51] which takes broader than geological views on stratification issues, or The Geological Society London (GSL) and the HKW, which repeatedly supported the AWG. The GSL facilitated the establishment of the AWG. The HKW promoted the interaction of the AWG with the cultural sphere. Initially, the HKW hosted and financially supported the AWG's first formal ('inaugural') meeting[52] in 2014 and the AWG meetings in 2021 and 2022, the latter as part of the GSSP project funded through the HKW; reported by the AWG:

> "thanks to the initiative of Berlin's *Haus der Kulturen der Welt* and its director, Bernd Scherer… This is yet another extraordinarily novel step for formal stratigraphy (the Anthropocene has provided quite a few, so far): for an arts/humanities organisation to fund such scientific work, and reflects the way that this chronostratigraphic assessment (even in its most drily technical aspects) has intersected with wider scholarship, and wider society….This new phase coincides with the structural change now due in the AWG, to conform with ICS rules." [AWG Newsletter 2019; p. 3].

The former director of the HKW Berlin, Bernd M. Scherer, conceived the HKW's Anthropocene Curriculum Project[53] (2013–2022). He is probably one of the key external players shaping AWG's broader environment. Together with Paul Crutzen (MPI Chemie, Mainz), these two persons of a Germany-based 'intelligentsia' were visibly driving the Anthropocene concept. Helmuth Trischler (*Deutsches Museum, München*) could likely be another person and German institution to mention in this context (Trischler 2016). Three newsletters of the AWG refer to the activities of the Deutsches Museum.

Support for meetings of university institutes (2015, 2019) or research institutes (2016, 2018), all but one located in Europe, has also benefited the AWG. Members of the AWG probably mediated the support. Finally, media and cultural organisations interested in the general issues addressed by the AWG have also been beneficial to the AWG's working environment (Sklair 2021). Members of the AWG responded to this interest, acknowledging the post-normal features of modern science, as shown by the expanding section of the newsletter reporting about media events and confirmed by the annoyance of others (De Wever and Finney 2018, p. 255).

[50] 'The mistaken drive to define the ‚Anthropocene'as an officially recognized unit of the Geological Time Scale' [paper 2491, 35th IGC]

[51] e.g. Thijs van Kolfschoten (Secretary SQS 2001–2015, INQUA president since 2019)

[52] AWG Newsletter 2013–14 p. 3

[53] https://www.anthropocene-curriculum.org/contribution/anthropocene-curriculum-genealogy/

2.4.4 The IUGS

Neither the IUGS annual reports nor the IUGS's scientific journal, Episodes, served as a hub for debates about a geological Anthropocene; the journal remained nearly silent. In mid-2023, before the AWG submitted its findings formally in October 2023, Episodes listed eight publications about a geological Anthropocene, five published since 2021. Six publications question the concept of a geological Anthropocene.

The reader must consult the news section of the SQS site to find intra-organisational communications about the Anthropocene concept. Finally, only the ICS Subcommission Annual Report 2022 lists publications about candidate GSSP sites for an Anthropocene. Likewise, it mentions related funding and meetings. If seen from an organisational perspective, this publication policy aligns with the task attribution in the IUGS/ICS system, although it can be interpreted as a signal of non-commitment or disagreement.

The first substantial reference in the IUGS reports to the AWG's business, other than reflecting irritations regarding, for example, media attention, was made in 2021:

> "Geoscientists from the Subcommission on Quaternary Stratigraphy (SQS) in the collaborative project of the Anthropocene Working Group (AWG), *Haus der Kulturen der Welt* (HKW) and *Max-Planck-Institut für Wissenschaftsgeschichte* (MPIWG) are making progress towards identifying Anthropocene GSSP locations.... Progress in the formalisation of the GSSP was presented during a session at vEGU 2021[54] "The physical record of the Anthropocene in geological archives". "[IUGS Annual Report 2021, p. 14].

This statement could witness a substantial shift in the IUGS's attitude or reflect the AWG's extraordinary achievement concerning fundraising. The change seems consistent with the views of the IUGS president Jon Ludden (2020 to 2024), who closes his 2020 address to the IUGS:

> "Together, we need to refine and redevelop the IUGS to respond to what the public and governments now expect of the largest global union of geological sciences—the International Union of Geological Sciences', ... 'we find ourselves confronted with daunting tasks. The most important is climate change and the environmental and social adaptions associated with reducing carbon and other emissions, so that we can reach NetZero-C emissions as rapidly as possible" [IUGS Annual Report 2020; pp .9–8].

It is beyond the scope of this study to describe how the IUGS's understanding of societal contexts in the Earth Sciences has evolved since the turn of the century. The president's statement reflects ample scientific and public debates on planetary-scale anthropogenic change.

[54] Annual Assembly of the European Geosciences Union 2021 (held as virtual meeting).

2.4.5 *Meetings and Publications*

The AWG members participate in scientific, public and media events and publish extensively. AWG's newsletters are not designed to systematically report AWG members' publications or participation, including scientific meetings/lectures or media events. Nonetheless, they provide some indication. In 2012, the AWG newsletter began to report on publications and media events; from 2014 onwards, the reporting was more detailed. However, it seems evident that some members of the AWG gave more detail or were more active than others. Some meetings and publications stand out because they trace the development of the AWG.

According to the AWG newsletter, in 2010–2012, about 10 scientific publications were published yearly; in 2013–2017, about 20; and in 2018–2021, about 40. The number of scientific conferences or lectures increased from 4 (2010–2012) to 20 (2013–2017) and about 70 (2018–2021). The number of reported media events increased from 2 (2010–2012) to some (2013–2017) and several tens (2018–2022). These numbers indicate a much-increased activity. They exclude the publications, meetings and media events prepared by the AWG. The number of media events suggests that AWG members cared about scientific exchange and public visibility, as detailed in Chap. 5.

Meetings

Between 2014 and 2022, the AWG held seven formal meetings. Six meetings were held in Europe, one in North America and none in other parts of the World. No formal AWG meeting was held during the first 4 years of the working group. All AWG meetings had external sponsors. The HKW, the most active sponsor, hosted three meetings (2014, 2021, 2022) in Berlin. In addition, gatherings of (some) members of the AWG likely happened in the margin of scientific or public events without specific reporting on them.

Some events related to the AWG's work before the first formal meeting in 2014 can be identified.

- Before setting up the AWG, the American Geoscience Union's fall meeting in 2008 included a session called 'Earth System Science for the Anthropocene', at which British stratigraphers presented their findings ('Stratigraphy of the Anthropocene'[55]) aggregated in the follow-up of the Burlington House (GSL premises) meeting[56] in 2006 and published 2008 in a journal of the Geological Society of America (Zalasiewicz et al. 2008).

[55] https://ui.adsabs.harvard.edu/abs/2008AGUFMGC11A0664Z/abstract

[56] https://www.geolsoc.org.uk/Geoscientist/Archive/December-January-2017-18/The-Geological-Anthropocene

- On 11 May 2011, the Geological Society London held a meeting on the Anthropocene[57,58,59] (convened by AWG member Mike Ellis), at which several members of the AWG, including Paul Crutzen, lectured. The meeting occurred after the Philosophical Transactions of the Royal Society published the theme-issue 'The Anthropocene: a new epoch of geological time?' edited by Mark Williams, Jan Zalasiewicz, Alan Haywood and Mike Ellis.
- In the same month, the third Nobel Laureate Symposium on Global Sustainability (16–19 May 2011, Stockholm) published 'The Stockholm Memorandum Tipping the Scales towards Sustainability',[60] co-signed by twenty Nobel Laureates, including Paul Crutzen, mentioning [p. 3],
- "[H]humans are now the most significant driver of global change, propelling the planet into a new geological epoch, the Anthropocene. We can no longer exclude the possibility that our collective actions will trigger tipping points, risking abrupt and irreversible consequences for human communities and ecological systems."
- On 1–7 July 2013, the First International Congress on Stratigraphy[61] was convened in Lisbon under the auspice of the ICS. It included a full-day special session, 'The Quaternary System and its Formal Subdivision', convened by Martin J. Head (Chair, SQS), Philip L. Gibbard (President, SACOM-INQUA) and Thijs van Kolfschoten (Secretary SQS 2001–2015, INQUA president since 2019). Some presentations, including Jan Zalasiewicz speaking about the onset of the Anthropocene in the mid-twentieth century, related to the work of the AWG. Subsequently, Martin J. Head, Philip L. Gibbard and Thijs van Kolfschoten summarised in two publications the discussion at the First International Congress on Stratigraphy on the Quaternary (Head et al. 2013; Head and Gibbard 2015); the latter including a detailed account of open issues related to a geological epoch Anthropocene at the end of the Quaternary.

These meetings seem central to shaping the AWG's views on possible features of a geological Anthropocene. The First International Congress on Stratigraphy and the subsequent publications can be understood as preparing the geological community to discuss the notion of a geological Anthropocene at the 35th IGC convened in 2016 in Cape Town.[62]

---

[57] https://www.nature.com/articles/473254a

[58] https://www.economist.com/briefing/2011/05/26/a-man-made-world

[59] https://www.science.org/doi/10.1126/science.334.6052.32

[60] http://globalsymposium2011.org/wp-content/uploads/2011/07/memorandum-signed.pdf

[61] Contributions of the AWG to the subsequent International Congresses on Stratigraphy and participation at events organised by the INQUA are mentioned in the AWG's reporting.

[62] One co-author of the book participated at the conference; see: (Bohle, M. & Ellis, E. C., "Ethical Requirements for Applied Earth System Science and Engineering". in 35th International Geological Conference, Geoscience for Society, Global Geoscience Professionalism and Geoethics (2016)).

The 35th IGC 2016 in Cape Town was a critical event in the development of the AWG. Three AWG meetings (October 2014, November 2015 and April 2016) preceded the 35th IGC, and four followed (September 2018, November 2019, September 2021 and May 2022). The 36th IGC scheduled for 2020 was massively truncated because of the coronavirus pandemic and, therefore, could not give a forum to discuss the work and findings of the AWG.

The 35th International Geological Congress

The 35th IGC, held in Cape Town in August/September 2016, was a remarkable event in terms of what was discussed (or not).

- First, the Anthropocene was only a minor subject in the vast public programme of the 35th IGC. The initially planned symposium, 'Earth Systems in the Anthropocene',[63] did not materialise. Instead, three contributions were made in session T47.1, 'Anthropocene: Utility and Definition', two by members of the AWG and one by Stan Finney, who refuted the need for a geological Anthropocene. In addition, six contributions using the term Anthropocene were made in other congress sessions.[64]
- Other than the 35th IGC's scientific programme, the online congress report[65] mentions,

> "Considerable media interest was generated from around the World on discussions relating to the identification of a new geological age, the "Anthropocene"—these discussions were held under the auspices of the Phanerozoic Earth History, Stratigraphy and the Geologic Time Scale theme and relate to the debate around the need to recognise a new geological epoch that is defined as having been significantly influenced by mankind."

- Next, the archive on the ICS website holds one report of a business meeting, namely the ICS report from its business meeting in the margins of the conference[66]:

> "Discussions then turned to the question of the possible definition of an "Anthropocene" time unit. MH [Martin Head] noted that a proposal for formal definition could be ready in

[63] http://www.35igc.org/SymposiumDetail/69/Earth-Systems-in-the- (proposed session abstract: 'Human activity has had a major impact on the workings of all parts of Earth's system, including sedimentary processes and dynamics of mountains, coasts and rivers; soils and biochemical cycles; mining and extractive industries; groundwater; and geohazards. This session considers the past, present and future patterns, processes and impacts of human (anthropogenic) impacts on different elements within the Earth's system; how such impacts can be identified from geologic records; and the role of human activity in changing the workings and management of Earth systems. A particular emphasis in this session is on the workings of Earth surface systems and processes in different physical environments on local to global scales.')

[64] Abstracts of 35th IGC: https://www.americangeosciences.org/information/igc

[65] http://www.35igc.org/Verso/5/Scientific-Programme

[66] https://stratigraphy.org/files/ICS-Business-IGC35.pdf

c. 2-3 years time, the preceding period being an establishment of the narrative concerning this possible division…." [2§, 4th paragraph]

This unique public publication of an ICS business meeting indicates its relevance.

- Finally, the outgoing IUGS president, Robert Oberhänsli, refers in the 2016 IUGS report to tensions at the 35th IGC:

"Remarkable problems were generated by single Participants and especially by a group working on latest Quaternary stratigraphy and promoting the term anthropocene [no capital in the text] in a strange way"[IUGS Annual Report 2016, p. 6].

Shortly after the 35th IGC, some AWG members gathered with peers at an extraordinary meeting on 17th March 2017 at the

"Max Planck Institute (MPI) for Chemistry in Mainz (Germany), hosted by Paul Crutzen on 16th March 2017, with participation of AWG members from University of Leicester (Jan Zalasiewicz, Colin Waters and Mark Williams) and those based in Germany (Paul Crutzen and Reinhold Leinfelder) and Norway (Davor Vidas) along with interested parties the MPI Chemistry (Ulrich Pöschl, Jos Lelieveld, Susanne Benner, Bettina Weber and Astrid Kaltenbach), MPI for the History of Science, Berlin (Jürgen Renn), Haus der Kulturen der Welt (HKW), Berlin (Bernd Scherer and Christoph Rosol) and IASS in Potsdam (Mark Lawrence and Franz Mauelshagen) to discuss future funding initiatives. In a follow up of this meeting, a project proposal was submitted to the Belmont Forum, which included a costed programme of multi-proxy analysis of candidate GSSP localities, but was eventually unsuccessful on the ground of formal aspects (eligibility)" [AWG Newsletter 2016-2017; p. 5].

After the 35th IGC, it took 2 years to convene a formal meeting of the AWG. It was held in September 2018 in Mainz at the premises of the MPI for Chemistry; for example, compliance with the ICS's requirement to identify a GSSP[67] was discussed. Likely (Damianos 2021; pp. 227–228), the meeting inspired Bernd Scherer (then director HKW) to innovate granting for the AWG and providing the HKW with a unique artefact (GSSP cores) exhibiting the tentative base of the Anthropocene. Before the AWG meeting in Mainz, the IUGS had accepted the GSSP for subdivision of the Holocene, which likely influenced the methodological choices of the AWG.

Publications

The work of the AWG was organised along publication projects, e.g. the Royal Society publication and the GSL book, which is a standard approach to structuring a scientific debate. For example, from 2009 to 2011, the AWG newsletter listed several publications, such as the articles in the Philosophical Transactions of the Royal Society 369A (Zalasiewicz et al. 2011). This theme issue of the Philosophical Transactions of the Royal Society was agreed upon in 2009, in conjunction with setting up the AWG.

---

[67] The ICS annual report for 2022 [p. 2] indicates the shift towards GSSP in the methodology preferred by the IUGS.

The GSL special publication 'A Stratigraphic Basis for the Anthropocene', published online in 2013 and printed in 2014, was also agreed upon when setting up the AWG. The three opening chapters of the book, with the lead authors J. Zalasiewicz, S. Finney and P. Gibbard, sketch the lines of discussion for the following years, including the event v. epoch debate from 2021 onwards (Bauer et al. 2021).

The leading publication of the period 2016–2018 was the book 'The Anthropocene as a Geological Time Unit' (Zalasiewicz et al. 2019), which records the state of the art of the Anthropocene as a geological concept, as consolidated by the vote of the AWG in May 2019. The book was conceived in 2016 before the 35th IGC. It was submitted to the publisher in autumn 2017 (a year after the 35th IGC) and was published (online) in February 2019, shortly before the formal voting of the AWG took place.

Among the AWG's scientific papers published in the same period, three can be singled out, first by authorship comprising most AWG members and second by the respective subject:

- 'The Anthropocene is functionally and stratigraphically distinct from the Holocene' (Waters et al. 2016) was published in January 2016;
- 'Making the case for a formal Anthropocene Epoch: an analysis of ongoing critiques' was published in April 2017;
- 'The Working Group on the Anthropocene: Summary of evidence and interim recommendation' (Zalasiewicz et al. 2017a) was published in September 2017.

The sequence of these three articles presents a claim, analyses the critics of the claim and sketches a consolidated position. It should be noted that Phil Gibbard, a member of the AWG from the beginning, is not one of the respective co-authors of these articles. Likewise, he takes a role in the book 'The Anthropocene as a Geological Time Unit', which is limited to an introduction about methodological issues, possibly indicating dissenting views.

The leading publication concluding the work of the AWG is the special issue of 'The Anthropocene Review' (published since the end of 2022) describing the candidate sites for the Anthropocene GSSP (Waters et al. 2023) before the AWG determined in July 2023 the prime candidate for its GSSP proposal. These publications are the primary input to subsequent discussions about the fate of the Anthropocene proposal, including the ramifications of

> "geological evidence for the Anthropocene and its repercussion for politics, society, the changing structures of knowledge creation, and, indeed, consensus-making." (Rosol et al. 2023; p. 228).

2.5 Featuring the AWG: Two Perspectives

The AWG's cultural skewness, maleness and European-centric composition are aforementioned and analysed in Chap. 3. Although not surprising, these features illustrate the AWG's embedding in an established framework of geological INGOs.

Despite what the AWG's name, Working Group on the Anthropocene, may convey to the reader, the working group is concerned with the geological status of the notion of Anthropocene. The following subsections also illustrate how IUGS and ICS insist on this feature.

2.5.1 A Bureaucratic Body

The bureaucratic features of the AWG are grounded in time-limited mandates, re-establishment processes and voting procedures for bodies of the ICS (Damianos 2021, p. 90ff.). These statutory rules describe an environment designed to settle methodological/scientific questions bindingly, at least accepted by most of the professional/disciplinary community. For example, the ICS statutes specify:

- "The chairs of the Subcommissions and the members of the Executive Committee together form the Voting Commission of ICS that votes on the formal decisions". [ICS statutes, §3];
- "The voting membership of the Subcommission consists of its management team together with up to twenty members, and is referred to as the Voting Subcommission... Subcommissions may appoint a reasonable number of corresponding members to advise voting members in achieving the assigned scientific tasks, e.g., participating in boundary working groups. The corresponding membership shall reflect regional and methodological diversity in an appropriate manner". [ICS statutes, §5.1].
- "Voting members of Task Groups are elected by its executive, in consultation with existing voting members, and confirmed by the management or executive of the ICS body under which the Task Group resides". [ICS statutes, §9.6]
- "The terms of office for the voting members of Subcommissions and Task Groups shall be the period between two IGCs, normally four (4) years, and can be extended for a maximum of two additional four (4) year periods". [ICS statutes, §9.2].

Further statutory provisions specify a supermajority (>60%), quorum (>60%), time delays and balloting procedures.

The AWG's newsletters first mentioned voting membership as distinct from advisory membership in 2019 when the AWG was re-composed after having taken a binding vote recommending a geological epoch called Anthropocene. To decide on the candidate GSSP, the AWG should re-constitute "into a voting membership with specialist expertise in chronostratigraphy to construct a formal proposal potentially acceptable to the ICS, and an advisory membership (in essence, our existing multidisciplinary membership)" [AWG Newsletter 2019; p. 3]. The re-composed AWG is mandated to propose candidate GSSPs; the proposal was made in 2023 [AWG Newsletter 2023, pp. 7–9].

The sequence of voting processes of the AWG, the SQS and the ICS precede a decision by the IUGS Executive Commission. The latter is "based primarily on the assurance that ICS policies and procedures were followed" (Finney 2014, p. 23). The former decisions, each a go/no-go step of an increasingly more general body, regard the acceptance of the geological epoch Anthropocene. This multi-stage decision process is designed for scientific thoroughness, bottom-up initiative, ample peer consultation and top-down control of methodological consistency.

The final decision of the IUGS is bureaucratic-political, namely confirming adherence to established procedures. Given this specific function of the IUGS, the epoch versus event debate taken up in the IUGS journal, Episodes, can be seen as addressing the concern of the IUGS/ICS executive that "the task group must give careful consideration to all positions and arguments" (Finney 2014, p. 24).

The AWG has voted twice, in May 2016 and May 2019, on the conceptual question of whether to propose a geological epoch, Anthropocene. The second (binding) vote confirms (a supermajority of 88%) the tentative views taken in 2016: 'the AWG have completed a binding vote to affirm some of the key questions that were voted on and agreed at the IGC Cape Town meeting in 2016'.[68] The AWG stated in May 2019 that the Anthropocene should be a formal chrono-stratigraphic unit defined by a GSSP of the GTS with a lower boundary around the mid-twentieth century.[69] Twenty-nine members of the AWG supported these views, four disagreed and none abstained. The subsequent voting in 2022 and 2023 on the details of the AWG proposal, i.e. the specific features of the proposed GSSP [AWG Newsletter 2023, pp. 7–9].

Given these voting processes, three periods illustrate the impact of the ICS's formal requirements on AWG membership. In 2012, Colin Waters was appointed secretary of the AWG, marking the end of the initialisation of the AWG. In 2016, the 35th IGC took place, and Martin Head (chair of the SQS) joined the AWG, marking the beginning of tighter cooperation with ICS. The destiny of the AWG members joining in 2009–2010 (initial configuration), 2011–2015 (working towards informal vote) and 2016–2021 (working towards candidates GSSP) regarding their likely part in voting on the GSSP proposal was:

- Nineteen members joined the AWG in 2009/10, nine left before 2021, six got advisory status in 2021, and four are voting members in 2021, of which one (Ph. Gibbard) left before the final votes on the GSSP.
- Twenty-two members joined the AWG in 2011/15, of whom four left in 2021, eight got advisory status, and ten are voting members.
- Ten members joined the AWG in 2016/21, one with advisory status, and nine are voting members. Six of these members joined after the binding vote in 2019, consolidating the AWG's expertise in chronostratigraphy.

[68] http://quaternary.stratigraphy.org/working-groups/anthropocene (accessed 16. February 2024; site reorganised later)

[69] The AWG published the outcome of the binding vote in its 2019 newsletter, which also reports the composition of the AWG in 2019. The AWG did not publish an explicit list of members participating in the vote, which can be reconstructed; the AWG had 34 members in 2019. 33 of 34 votes were casted; the author assumes that Paul Crutzen (honorary member) did not vote. The SQS statement gives the composition of the AWG at date of publishing in 2020 and not at the date of the vote.

The re-composition of the AWG focussed AWG's scientific profile on expertise to assess the geological validity of the proposed candidate GSSP by adding new members and distinguishing advisory and voting members. These changes happened after the AWG voted to favour a geological epoch, the Anthropocene. The six voting members joining the AWG after 2019 and those voting AWG members who are opposing a geological epoch Anthropocene together have a blocking minority stake in the re-composed AWG, i.e. is sufficient to block the required supermajority for a GSSP as the base of a geological epoch Anthropocene. Hence, repeating the 2019 vote by presenting a candidate GSSP would not be possible for the AWG. Hence, the consensus finding within the AWG took a third iteration.

The position of the AWG in 2019 was more affirmative than in 2016 when the AWG prepared input to the 35th IGC at the

> "third meeting of the AWG, coordinated by Davor Vidas, was held in Oslo on 22nd–23rd April 2016, kindly hosted and financially supported by the Fridtjof Nansen Institute (FNI) and Research Council of Norway. This AWG meeting, with Martin Head (current Chair of SQS) and Felix Gradstein (former Chair of ICS[70]) as invited speakers and attendees, focussed on a consensus statement for IGC2016, notably key signals and potential environments which could host a GSSP." [AWG Newsletter 2016-17, p. 4].

The AWG never published the outcome of the 2016 vote in its newsletter. It can be found in the media note of Leicester University on 29th August 2016[71] (issued some days after the start of the 35th IGC), mentioned in publications (Zalasiewicz et al. 2017a) and was summarised in the information published in 2020 by SQS[72] on the binding AWG vote.

With the binding vote in 2019, the SQS published a statement entitled 'What is the Anthropocene?—current definition and status', which sketches what may shape upcoming debates. Also, it centres the Anthropocene question on system change ("taking us out of the Earth System state typical of the Holocene Epoch"), re-iterates the mandate of the AWG ("we examine the Anthropocene as a geological time (chronostratigraphic) unit and potential addition to the Geological Time Scale"), addresses potential markers ("The sharpest and most globally synchronous of these signals, that may form a primary marker, is made by the artificial radionuclides spread worldwide by the thermonuclear bomb tests from the early 1950s").

Furthermore, and essential for several contentious issues, the statement affirmatively elaborates on contexts of a geological notion of Anthropocene, namely that it should be "scientifically justified" and "useful as a formal term to the scientific community". Also, it is conceded that the value of the notion beyond global change and ESS communities needs discussion. Informal use of an Anthropocene in non-chronostratigraphic contexts, such as the diachronous human impact on the planet,

---

[70] ICS chair 2000–2008 succeeded by Stan Finney from 2009 onward.

[71] Media Note [Posted by ap507 at Aug 29, 2016 01:11 PM], Accessed and downloaded 18th October 2021, not accessible in 2022 and later, secondary source: Damian Carrington, The Guardian, 29 Aug. 2016, https://www.theguardian.com/environment/2016/aug/29/declare-anthropocene-epoch-experts-urge-geological-congress-human-impact-earth

[72] http://quaternary.stratigraphy.org/working-groups/anthropocene (accessed 16. February 2024; site reorganised later)

is also acknowledged, referring to lithostratigraphy as the appropriate geological approach. Unsurprisingly, publication activity in consecutive years addresses these issues (Gibbard et al. 2021; Head et al. 2022; Waters et al. 2022; Koster et al. 2023), for example, in the form of the event versus epoch debate, which picks up the question of the scientific utility of the epoch notion.

2.5.2 *The Timeline*

The work of British stratigraphers before 2009, including publication activities, resembled a (national) proto-AWG, which led to the SQS setting up the AWG. The AWG provided an opportunity to continue personal involvement. More than half of the AWG members were affiliated with institutions in the United Kingdom.

Formally consolidated as a working group, the membership of the AWG in 2010/11 increased (by 50%) and diversified. These consolidation and diversification processes continued in 2012 and 2013 in junction with visible (international, interdisciplinary) scientific conferences, an unsuspicious 34th IGC[73] (Brisbane) in 2012 and the first formal meeting of the AWG in October 2014 after the statutory extension of the AWG's existence in 2013 according to the rules of the ICS. At that time, the AWG had 35 members, counting the departure of three founding members. Hence, about two-thirds of AWG members joined the working group when activities were centred on scientific publications, indicating that AWG is not a closed club.

The preparation for the 35th IGC (2016), the participation at this event and the recovery post-event seem to englobe

> "something of a transition interval for the AWG—negotiated, we hope, reasonably successfully. The end of one era for the AWG—if one may put it that way—came with the International Geological Congress at Cape Town in August 2016,... ; [the Anthropocene] should be defined by a GSSP ('golden spike') rather than a GSSA[74]; and that—as a consequence—we should proceed towards building a formal proposal to forward for consideration, initially, to the Subcommission on Quaternary Stratigraphy." [Jan Zalasiewicz, AWG report 2016-17, p. 3].

During this transition period, membership in the AWG was renewed further (by a little less than 20%). The AWG was formally reinstalled by the chair of the SQS (Martin Head),[75] who could do so, given the statutory provisions of the ICS. Also, Martin Head joined the AWG in the reporting period 2016–2017, considering the wording of the AWG newsletter 2016–2017, likely from 2017 onwards. However,

---

[73] 'In Brisbane, ...Media interest in the 'Anthropocene' kept the Chairman busy responding to interview requests'. [IUGS Annual Report 2012, p. 13; sole reference to Anthropocene, https://www.iugs.org/_files/ugd/f1fc07_ae73cfdf779549509c61c78ad483c20b.pdf]

[74] Global Standard Stratigraphic Age (GSSA)

[75] §7.1 of the ICS statute states that a task/working group '*should be dissolved*' after 8 years although the chair of a subcommission has the discretion to reconvene it.

he had participated in the AWG meeting in April 2016, which prepared for the 35th IGC, among other tasks.

In subsequent years (2018–2022), the work of the AWG focused on consolidating the proposal of the Anthropocene being a geological epoch defined by a GSSP after the option to use a GSSA has been abandoned in the context of the 35th IGC. The voting took place in May 2019.[76] Subsequently, the AWG was renewed; in Jan Zalasiewicz's words:

> "This new phase coincides with the structural change now due in the AWG, to conform with ICS rules.... should allow both the technical side of our work to proceed in classical manner, and for the range, spirit, scholarship and energy of our collective endeavours to continue, I hope, undimmed." [AWG Newsletter 2019, p. 3].

In 2020, the AWG membership was re-casted as voting and advisory members. The voting members are bound by the provisions of the ICS/IUGS and assume the task of proposing a GSSP. In 2020, the functions of the chair and secretary of the AWG were re-assigned. Jan Zalasiewicz took the chair of the SQS, the ICS subcommission to which the AWG submits its proposal of a GSSP for the Anthropocene.

The COVID-19 pandemic slowed the activities of the AWG, the ICS and the IUGS, including delaying the 36th IGC (New Delhi, India) by about 2 years to March 2022. Four members of the AWG[77] submitted a contribution, 'Progress in investigating potential Global Boundary Stratotype Section and Point (GSSP) candidates for the Anthropocene Series' [abstract 4775–4553][78], outlining the ongoing work of determining a suitable GSSP. A further contribution by Martin Head on the Quaternary states,

> "The Anthropocene is presently undefined but as currently envisioned would start ~1950 having the rank of series/epoch, and terminate the Holocene without interfering with its subdivision other than to end the Meghalayan." [abstract 3848-3242].[79]

The second part of this assertion is far from innocent because agreeing on the subdivision of the Holocene within the IUGS framework[80] ended a lasting quarrel in the geological community and between geological INGOs.

Furthermore, the decision about the Holocene subdivision frames the work of the AWG, as summarised (Walker et al. 2018, p. 8):

> "Whereas most current GSSPs are based upon biostratigraphical evidence as their primary markers, it has been recommended that future GSSPs should have physico-chemical markers as an integral part of their guiding criteria... The definitions here follow that recommendation, as all three of the GSSPs are defined on the basis of significant physical and chemical markers. They reflect abrupt climatic events... Indeed, the designated GSSPs are the best-resolved, both stratigraphically and temporally, within the entire geological time

---

[76] SQS report on the AWG's vote in 2019: http://quaternary.stratigraphy.org/working-groups/anthropocene (accessed 16. February 2024; site reorganised later)

[77] Neil Waters, Simon Turner, Martin Head, Jan Zalasiewicz

[78] https://36igc-virtual.in/abstracts/

[79] https://36igc-virtual.in/abstracts/

[80] http://quaternary.stratigraphy.org/iugs-ratifies-holocene-subdivision/ (accessed 16. February 2024; site reorganised later)

> scale. They… provide stable points of reference for Holocene stages/ages and subseries/subepochs into the future".

Hence, these decisions of the ICS, endorsed by the IUGS, namely to subdivide the Holocene using physico-chemical markers, give a foundation to base the GSSP for the Anthropocene on physico-chemical markers. Furthermore, the age of Holocene subdivisions is counted from the year 1950.[81] Hence, setting the base of the Anthropocene around the mid-twentieth century would not question the age of the Holocene subdivisions.

2.6 Staging the AWG

2.6.1 Conceiving

The AWG's 'prenatal' stage, situated before formally establishing the working group, relates to a GSL's Stratigraphy Commission meeting. The Burlington House Meeting of 17th May 2006 took place several years before the formal inauguration of the AWG and 6 years after the IGBP meeting at which Paul Crutzen suggested ending to call the recent historical past, the present and future times '*Holocene'*.

We know about the Burlington House Meeting only through a short publication in Geoscientist Online (Zalasiewicz et al. 2017b) published in 2017, stating[82]:

> "One of us (JZ), as its (then) Chair, introduced a draft manuscript on the Anthropocene for consideration as a Commission publication… The Stratigraphy Commission [of the GSL] is not an international body, so cannot "adjudicate"—though it can pass comment. Perhaps somewhat to its own surprise, it agreed (by large majority) that the Anthropocene seemed "geologically reasonable" and might be considered for formalisation."

Following this somewhat forgotten meeting, many members of the GSL's Stratigraphy Commission jointly published a paper in the journal of the Geological Society of America (Zalasiewicz et al. 2008), one of several ten indexed papers on the Anthropocene published yearly between 2006 and 2008 (Brauch 2021, Table 22.1 p. 313). Also, a session at the American Geosciences Union fall meeting in 2008, 'Earth System Science for the Anthropocene',[83] occurred. Prepared in this manner, the SQS (chair Phil Gibbard) of the ICS (chair Stan Finney) installed the AWG in the subsequent year and 'appointed' its chair (Jan Zalasiewicz), who was deemed qualified, having led the subcommission of the ICS looking after methodological questions.

[81] The 2020 edition of the GTS moved the reference date to year 2000.

[82] https://www.geolsoc.org.uk/Geoscientist/Archive/December-January-2017-18/The-Geological-Anthropocene; see subsection "The dawn of the AWG" for a more detailed quote.

[83] Blog post by Erle Ellis https://ecotope.org/blog/pushing-back-the-anthropocene-at-the-agu/

2.6.2 *Establishing*

The AWG's setting up, composition and initial activities happened in 2009–2011/12. The records of the ICS and the IUGS of 2008 and 2009 (annual reports) do not provide information on the specific reasons or motives for installing the AWG. The momentum of the scientific debate on a geological epoch called the Anthropocene was probably sufficient; a risk was perceived that extra-geological scientific communities were establishing a factual terminology in an area where geologists claim epistemic leadership. As a correlative can be assumed, the British stratigraphers accumulated evidence for a geological foundation of the Anthropocene through their recent activities to revise the stratigraphy for the Quaternary of the British Isles. Given these circumstances and considering that most responsible officers were British and that a subcommission of the ICS has some autonomy to establish working parties, the choice to install a working party of the SQS seems obvious.

The timing of the decision seems suitable because the IUGS and the INQUA, first, had settled the quarrel about the Quaternary, and second, the findings on the subdivisions of the Holocene were pending. In that context, exploring a possible stratigraphic foundation of an Anthropocene (and its lower boundary) made sense. The initial composition of the AWG reflects concerns about analysing the question of a geological epoch, the Anthropocene, being a broader concept than geological. Therefore, including scientists who had diverging opinions from the beginning (e.g. Phil Gibbard) and attempting some disciplinary diversity (e.g. Erle Ellis, Davor Vidas) was suitable.

The initial AWG work program was published in December 2009 in the newsletter [AWG Newsletter 2009, pp. 3–5], initialised by the phrasing:

> "In general, we have to critically compare the current degree and rate of environmental change, caused by anthropogenic processes, with the environmental perturbations of the geological past, to assess the scale of current global change in geological terms. The following parameters may fall within our area of interest:"

In subsequent paragraphs, these parameters are detailed, some broader (societal) considerations are made, target dates (ICS meeting in 2010, IGC 2012) are mentioned, and the operating mode of the AWG is sketched:

> "...in common with most other ICS working groups we do not have a budget, for travel or otherwise. Work will be largely conducted by e-mail, with occasional ad hoc meetings whenever a sufficient number (two or more) members find themselves at, say, the same conference. "Official" distribution of results, news and so on may best be done through Newsletters circulated to all of us at regular intervals—perhaps quarterly (and copied to the SQS, ICS and INQUA) with information collated centrally by Chair and Secretary. In addition, an annual report needs to go to the SQS/ICS."

The AWG work program published in 2010 is more focused and less ambitious regarding the time needed.

> "Once this work is achieved (realistically, over the next few if not several years), we may then be in a position to effectively consider the Anthropocene as a whole, and perhaps arrive

at some agreement on its possible status, hierarchical level and extent." [AWG Newsletter 2010, pp. 4–5].

In consecutive years, the newsletter section referring to the AWG work programme has been short; instead, the description of the workshops has taken up ample space.

In practice, the operating mode of the AWG during the period 2009–2011/12 was determined by 'publication projects undertaken in correspondence' because no resources were available to fund a meeting. Coupled distinctly with installing the AWG, two dedicated publication projects were prepared: a series of papers in a thematic issue of the Philosophical Transactions of the Royal Society, 'The Anthropocene: a new epoch of geological time?' compiled and edited by Mark Williams, Jan Zalasiewicz, Alan Haywood and Mike Ellis, published in 2011, and a book in the special publication series of the GSL, 'A Stratigraphic Basis for the Anthropocene', published online November 2013. These publication activities of the AWG and individual publications of AWG members occurred in a favourable environment in which several new dedicated journals emerged, such as 'The Anthropocene Review' by SAGE, 'Anthropocene' by Elsevier or 'Elementa' by the University of California Press.

The GSL book anticipated most debates of the consecutive years, including mandatory adherence to the established methodology of the GTS (Finney 2014) and the recent discussion on a geological Anthropocene event versus a geological Anthropocene epoch:

"As such, the activities of humans cannot then be used again to justify the definition of a discrete Anthropocene division. This is the view of many Holocene geoscientists and archaeologists, who do not accept that the Holocene has ended "(Gibbard and Walker 2014, p. 35).

The debates reflected in these publications, apparent from the AWG's conception onward, find correlatives in the meeting on the Anthropocene organised by the GSL (2011) and the Stockholm Memorandum of Nobel Laureates (2011).

Hence, during 2009–2011/12, the AWG initiated its activities within an ample publication activity that gained public visibility. The operation mode of the AWG was conventional and academic, although the scientific diversity of the AWG was vast compared to the customary IUGS/ICS practices. It reflected a vision of an Anthropocene notion much more comprehensive than being an element of the geological terminology. This scientific diversity, when seen from the outside, drove studies of inclusivity and evidence-finding characterising the AWG (Lundershausen 2018b, 2018c; Will 2021) as detailed in Chap. 4. The range of considered scientific approaches, GSSA, GSSP or techno-fossils, was broad. Also, various dates of onset (base) within recent historical times were considered (Zalasiewicz et al. 2017a).

2.6.3 First Metamorphose

The period 2013–2017 englobes the statutory re-conduct of the AWG after 4 years of existence. In this period fall the first three plenary meetings of the AWG, the structured discussions of the Anthropocene concept at two crucial meetings of the geological community (The first International Congress on Stratigraphy, July 2013; the 35th IGC 2016, August/September 2016) and the impromptu meeting of some AWG members and third parties in March 2017 at the MPICh (Mainz). Regarding the AWG's publication projects, which structure the work, the initial draft of the book 'The Anthropocene as a Geological Time Unit' was prepared.

Four scientific AWG publications indicate how thinking evolved. The articles: 'When did the Anthropocene begin? A mid-twentieth century boundary level is stratigraphically optimal' (Zalasiewicz et al. 2015), 'The Anthropocene is functionally and stratigraphically distinct from the Holocene' (Waters et al. 2016), 'Making the case for a formal Anthropocene Epoch: an analysis of ongoing critiques' and 'The Working Group on the Anthropocene: Summary of evidence and interim recommendation' (Zalasiewicz et al. 2017a) describe a shift away from being explicitly affirmative. Towards the end of the period, and in line with the ICS practices, the preference for a GSSP was affirmed. A shift in the preferred geological methodology appears to have occurred.

These shifts of attitudes towards the AWG were caused by events in 2016, which the AWG chair describes as "something of a transition interval for the AWG—negotiated, we hope, reasonably successfully…" [AWG Newsletter 2016–17, p. 3]. This statement reflects the fact that the AWG could not convince its peers in Geology at the 35th IGC, starting with failing to convene a dedicated scientific session in the conference programme and concluding in a crushingly laconic way during the ICS business meeting [§2, 4th paragraph][84] "…the preceding period being an establishment of the narrative concerning this possible division… The role of Martin Head (chair of the SQS since 2012) was likely pivotal because the 35th IGC was a place of irritation,[85] and the AWG witnessed 7 years of work being qualified in the public report of an ICS business meeting as narrative. Likely, this qualification was the ticket to gain a second life by continuing the working group for some years beyond the habitual statutory lifetime of 8 years in stricter cooperation with the ICS and accepting to replace the 2016 vote with a 'binding' poll in 2019, which turned out to be more affirmative than the vote in 2016.

Driven by the 35th IGC, including the increased 'insistence' on using a GSSP instead of a GSSA for the lower boundary of the Anthropocene, the ICS requires involvement in identifying the GSSP upstream to the formal proposals and voting.

> "An AWG-SQS-sponsored open workshop on the Anthropocene to be organised at which the SQS voting membership and the executive of ICS will be encouraged to attend along

[84] https://stratigraphy.org/files/ICS-Business-IGC35.pdf

[85] https://www.iugs.org/_files/ugd/f1fc07_4e5e716ce9a14eb290219a16bac40741.pdf

with the AWG membership. The focus will be on the stratigraphic expression of the Anthropocene and progress made on selected sections "[ICS Annual Report 2017, §9].

The ICS reports of the following years do not refer to this specific meeting. However, the fourth AWG meeting in September 2018 (Mainz, Germany) may qualify as it explicitly tackled ICS business', 'SQS progress for Holocene stage definitions (Martin Head)' and 'ICS protocols and key concerns regarding formalisation (Phil Gibbard)' [AWG Newsletter 2018, pp. 4–5]. Similarly, the fifth AWG meeting in November 2019 (New Orleans, USA) included sessions like 'The process involved in seeking formalisation' [AWG Newsletter 2019, p. 8]. Finally, in 2022, a meeting in Florence (AnthroFlor) was scheduled by the SQS[86], serving the intended purpose partially (Koster 2022).

2.6.4 *Second Metamorphose*

Three events marked the fourth period of AWG development (2018 to 2023). First, the 'binding vote' of the AWG took place in March–May 2019. Second, the GSSP project of the HKW started in May 2019 after the HKW secured funding in 2018 [AWG Newsletter 2018, p. 3]. Third, the statutory renewal of the AWG happened at the end of 2019. Likewise, this period was impacted by the Coronavirus pandemic (2020/21), which also slowed down the IUGS and ICS activities.[87]

During the 2018–2023 period, four meetings of the AWG were scheduled, two before the pandemic (September 2018, November 2019) and two since September 2021, all with the primary task of finding candidates for the Anthropocene GSSP stratotypes. The multi-year cooperation between the AWG and the HKW is embedded into a vast cultural programme of the HKW[88]:

"How does the global fingerprint of humankind manifest itself in Earth's sediments? ...Since 2019, the Anthropocene Working Group has been searching for stratigraphic evidence indicating the geological onset of the Anthropocene at various sites around the world. Through a close reading and contextualization of this material evidence of planetary upheaval, the yearlong program Evidence & Experiment gathers material testimonies of the Anthropocene stratum. ...It is this uncharted space between the evident and the experimental, between what is "given" and the collective search for adequate responses, that the program explores through a series of events, online publications and an exhibition."

The 36th IGC, scheduled for February 2020 in New Delhi, got delayed by 2 years because of the Corona pandemic and was heavily truncated. Finally, it did not serve as the venue for the business meetings of the IUGS and its constituent bodies. Subsequently, the 36th IGC could not be an opportunity to discuss the 'binding vote' of the AWG (May 2019), following what was agreed at

[86] https://stratigraphy.org/files/ICS_SubcommReport2022.pdf

[87] See, for example, the IUGS Annual Report 2020 report [p. 10–11].

[88] https://www.hkw.de/en/programm/projekte/2022/evidenz_experiment/start.php

> "the Mainz AWG meeting in September 2018... It is customary ICS procedure to formally decide on the approximate level of a boundary prior to the search for candidate GSSPs... The process commenced on 19th March 2019... The voting results, below, were announced on 21st May 2019". [AWG Newsletter 2019; p. 4].

The vote[89] confirmed that most AWG members supported an Anthropocene epoch beginning in the mid-twentieth century. The vote was taken when it was known that funding was available to help with the search for the GSSP. It is not easy to imagine how the poll and the search for GSSPs could have been undertaken if the HKW had not secured substantial funding for its Anthropocene Curriculum project.[90] The HKW used a significant part of the financing to obtain the cores holding the candidate GSSP for its exhibition through the AWG.

> "Thanks to the initiative of Berlin's *Haus der Kulturen der Welt* and its director, Bernd Scherer, the lengthy, complex and expensive process of selecting and analysing potential GSSP candidate sites, each represented by its own scientific team working in partnership with us, has begun in earnest, led by Colin Waters and Simon Turner. This is yet another extraordinarily novel step for formal stratigraphy (the Anthropocene has provided quite a few, so far): for an arts/humanities organisation to fund such scientific work, and reflects the way that this chronostratigraphic assessment (even in its most drily technical aspects) has intersected with wider scholarship, and wider society" [AWG Annual report 2019, p. 3].

A minority of AWG members took a dissenting position. Subsequently, Phil Gibbard (ICS secretary general, voting member of the ICS), the AWG advisory AWG members Erle Ellis and Mathew Edgeworth and other co-authors (all known to be critical to AWG's work) published an article in the IUGS journal, Episodes, favouring an 'Anthropocene Event' instead of a geological epoch (Gibbard et al. 2021). The authors argue that the former would favour scientific research, a view consistent with an earlier publication (Gibbard and Walker 2014, p. 35). The argument also corresponds to concerns mentioned in the qualifying statement accompanying the publication of the vote by the SQS.[91] Likely, this publication in, Episodes marks the watershed, proposing an alternative to the likely AWG proposal behind which opponents of the epoch concept could rally.

The significance of the extraordinary partnership between the AWG and the HKW (Damianos 2021) is on par with the recognition that the IPCC gave the Anthropocene concept, given that the IPCC is the intergovernmental scientific advisory body in the UN process on climate change negotiations. The IPCC used the Anthropocene concept in its 1.5 °C report,[92] including AWG publications as part of its scientific references (IPCC 2018). The AWG claims its role in this achievement [AWG Newsletter 2018, p. 4]. This international recognition beyond the scientific

[89] http://quaternary.stratigraphy.org/working-groups/anthropocene (accessed 16. February 2024; site reorganised later)

[90] https://www.anthropocene-curriculum.org

[91] http://quaternary.stratigraphy.org/working-groups/anthropocene/ (accessed 16. February 2024; site reorganised later)

[92] Special Report on the impacts of global warming of 1.5 °C above pre-industrial levels and related global greenhouse gas emission pathways, in the context of strengthening the global response to the threat of climate change, sustainable development, and efforts to eradicate poverty.

sphere and the exceptional financial support to run the GSSP project likely consolidated the (bargaining) position of the AWG with the IUGS/ICS framework. The IUGS refers to the AWG in its annual report 2021 [IUGS Annual Report 2021, p. 14]. It is unique that the IUGS notices a lower-ranked body of one of its constituent organisations in its annual report.

The formal proposal of the lower boundary of the geological epoch Anthropocene was made by the renewed AWG (composition of the year 2020) under the chairpersonship of Colin Waters. The annual newsletter of the AWG 2021 [p. 4] informs,

> "[t]he schedule of activities in 2022 related to completion of the GSSP work has also become clearer. The key meeting in our calendar will be the AWG meeting at HKW, Berlin on 18th to 19th May. AWG members will be invited to attend in person or remotely. The results of the studies will then be published as separate chapters within a special thematic issue of The Anthropocene Review to be edited by Waters, Turner, Zalasiewicz and Head.

The articles describing the GSSP began to be published in 'The Anthropocene Review' at the end of 2022 (Waters et al. 2023).

2.6.5 *The Final Stage*

The AWG's voting process for selecting the candidate GSSP and ancillary decisions requested by the ICS took its time. In July 2023, the candidate GSSP, Crawford Lake (Canada), was announced (Witze 2023). Subsequently, the AWG proposed to the SQS in October 2023 the Lake Crawford site in Canada to mark the base of the Anthropocene as a new epoch of the GTS, ending the Holocene, then finding the proposal rejected in March 2024 (Witze 2024b).[93]

The debate of previous years, whether geological event, geological episode or geological epoch (Bauer et al. 2021; Gibbard et al. 2021; Waters et al. 2022), foreshadowed the rejection decision. It requires further analysis of whether the AWG's message package of 'shift of Earth System dynamics; caused by affluent western societies; measured by the radioactive outfall of testing nuclear weapons' simply was too much, and, hence, denial was the default reaction, i.e. (Koster and Gibbard 2024) at an international conference,

> "[i]t views Geological Time Scale protocols as unsuitable for archaeological and contemporary developments, regards unemotive references to humanity's most abhorrent invention as distasteful, and visualizes the Anthropocene Event as valuably informing a new zeitgeist for our troubled world."

This submission of the proposal to SQS formally ends the lifespan of the AWG. The subsequent dealings of the SQS, the ICS and the IUGS are outside the scope of this chapter. They are addressed in Chap. 6, including issues like scientific reasons (preference of the event concept), procedures and practices. Considering the IUGS's decade-long silence regarding the Anthropocene question, the IUGS Annual Report

[93] During the year 2023 Erle Ellis, Ph. Gibbard, and Matt Edgeworth left the AWG.

2023 is surprisingly vocative [IUGS Annual Report 2023, pp. 6, 16, 32], foregrounding the looming decision.

The reasons for the rejection decisions spotlight cultural topics (Koster et al. 2023; Koster et al. 2024). In this perspective, the ICS/IUGS decision mirrors the broader Anthropocene questions discussed in Chap. 4. In its public statement (IUGS 2024), the IUGS sidelines scientific arguments by referring to publications and associates the organisation with the event concept. The scientific merits of the AWG members (see Chap. 3) are acknowledged. The IUGS statement portrays the current stage of the Anthropocene debate, a disagreement (Bohle 2023) between two economies of knowledge (Renn 2020) presented *ad coram publico,* which manifests the ample public outreach of the Anthropocene debate (see Chap. 5).

2.7 Discussion and Conclusions

2.7.1 Narrative: Bureaucratic Ways

The IUGS coordinates the international interaction of academic geological disciplines, which is a sizable endeavour in a diverse, mainly descriptive field (Baker 1999; Phillips 2012; Hazen 2012; Langmuir and Broecker 2012). One of the IUGS's constituent bodies, the ICS, coordinates peer agreement on essential methodological matters[94] and conservation of geological findings (Finney and Hilario 2018).

The IUGS/ICS system is a resource-poor, reputation-rich and primarily rule-based apparatus for finding consensus on intra-disciplinary issues, having limited experience in handling questions at science-public interfaces. The decision processes are based on majority consensus among peers and are taken at subsequent hierarchical levels. Working groups of ICS's subcommissions are an essential foundational element of this bureaucratic apparatus. Funding for activities of the ICS's working groups is scarce; essentially, it does not exist. The unpaid ICS and IUGS officers may access funding for committee meetings. In such circumstances, individuals in working groups use resources of their professional affiliations or projects to support the IUGS/ICS activities that align with professional contexts, including reputation building among academic peers.

These material and institutional circumstances, including established conventions and practices, determine the habits of the IUGS/ICS system and constrain the operational options of its officers and constituent bodies. However, these circumstances do not curtail intellectual (academic) options, which depend very much on reputation among peers,[95] such as options for mobilising resources 'in kind' or

[94] §2 Statute of the ICS 'ICS is a body of expert stratigraphers founded for the purpose of promoting and coordinating long-term international cooperation and establishing and maintaining standards in stratigraphy'. *https://stratigraphy.org/statutes*

[95] „I find it is something of a privilege to be able to pursue this [Anthropocene] science in the present company that is this working group"(Jan Zalasiewicz) [AWG Newsletter 2013–14; p. 2]

exercising convening power for joint publications. For example, the network and skills of leading individuals to acquire attractive publishing opportunities determine how successful the coordination could be. Attuned to such features, coordinating work through publications (and correspondence) is usual. The 2006 Burlington House meeting, which led to the emergence of the proto-AWG, already witnessed these features. Likewise, the activities of the AWG around publications confirm habits of an approach of 'coordination by publication'. Appointing Jan Zalasiewicz as chair of the AWG secured, from an IUGS/ICS perspective, a leadership that is experienced and well-networked within the ICS and has proven publication skills, convening power and good interpersonal skills. All these characteristics are needed to oversee the complex subject, namely whether a geological epoch Anthropocene in the GTS would be due.

Considering the functioning of working groups, group discussions in resource-limited academic environments mainly occur in the margins of conferences, which scientific organisations organise with respective (global) outreach. Dedicated workshops require additional resourcing, which could be found, for example, through scientific institutes that provide rooms and 'in kind' technical support. Individual travel would be funded through professional affiliations or projects. Conceptually, acquiring public or philanthropic funding for workshops would be possible.

The AWG newsletters show successfully coordinated activities using publications, participation in scientific conferences and workshops sponsored 'in kind'. For most of its lifespan, the AWG and the proto-AWG of 2006–2008 functioned in this manner, albeit with nuances. During the first years of the AWG, until the 34th IGC 2012 in Brisbane, the AWG worked frugally without dedicated group meetings. AWG members participated in national, regional and international conferences and sized opportunities for public visibility, for example, at the AGU Fall Meeting 2012.[96] In comparison, attendance at the 34th IGC in 2012 was minimal [AWG Newsletter 2012; p. 7]. In the following years, the plenary meetings of the AWG were hosted by third parties which sponsored the event.

The inaugural AWG meeting was hosted and supported by the HKW, which was embedded into a comprehensive programme of cultural activities[97] that continued for several weeks. The AWG's contributions at the inaugural meeting to the activities of the HKW were made in a half-day session, 'Human Impacts and Their Consequences: An open forum on the occasion of the first meeting of the Anthropocene Working Group' was moderated by the director of HKW, Bernd Scherer and the AWG chair Jan Zalasiewicz, concluding with a discussion of the

> "quality and magnitude of the shift posed by the Anthropocene, critical assessment of (science-)political solution pathways and a research agenda beyond sustainability, linking scientific practice with societal relevance and local to global strategies of knowledge production". [AWG Newsletter 2015; p. 6]

[96] Press conference: https://www.youtube.com/watch?v=RtgfgbHCNjY

[97] Online publication: https://issuu.com/hkwberlin/docs/anthropoceneproject_areport_booklet (see pages 70–77 for participation by the AWG)

The host and design of the inaugural meeting are exceptional for a geological working party and illustrate the prowess of the Anthropocene epoch concept to engage discussions across disciplines. Compared to the more conventional academic meetings in later years, the inaugural AWG meeting (2014) offered an unconventional embedding for an IUGS/ICS working group, which likely preconditioned that the AWG responded further to the broader contexts of the Anthropocene concept. Subsequently, the evolution of the AWG transgressed the usual perimeter of an IUGS/ICS working group limited to intra-geological concerns. The AWG's initial composition included expertise that was more comprehensive than that of geological stratigraphy, and this feature has already opened a broader perspective of the AWG on the Anthropocene. The AWG's second (2015) and third (2016) meetings focused on the publication activities and preparing AWG's contribution to the 35th IGC. They were hosted by academic institutions, exposing the AWG members to scientific disciplines other than Earth Sciences.

Over time, the publications and public interventions of the AWG members and reactions of peers within the IUGS/ICS system showed the strain under which a geological concept of an Anthropocene emerged. A rupture occurred during the 35th IGC. The peers of the AWG in the IUGS/ICS system called the findings of the AWG a 'narrative', disqualifying them in terms of geological stratigraphy. The AWG took about 2 years to recover from this verdict, mainly through a publication project focusing on geological stratigraphy (Zalasiewicz et al. 2019). The fourth (September 2018) and fifth (November 2019) AWG workshops were held in academic settings mainly dedicated to this publication, and the AWG's 'binding vote' was delivered in May 2019. The AWG confirmed the position taken previously in May 2016. Subsequently, the ICS profoundly reconfigured the AWG to assess possible candidate GSSP of a geological base of the Anthropocene, tracing the onset of the Great Acceleration. Probably, seen from a supposed IUGS/ICS perspective, the unruly behaviour of the 'early' AWG got tamed.

In the meantime, metaphorically phrased, 'the continental plates of the AWG sphere shifted'. Substantial intellectual and financial resources from third parties became available to the AWG in 2019. Subsequently, the AWG gained such a level of operational autonomy that its activity was reported in the annual report 2021 of the IUGS, which is pretty extraordinary.

The AWG's Anthropocene GSSP project started in 2019, benefitting from the modified methodological background provided by recent decisions of the IUGS/ICS. In 2018, the IUGS agreed to the Holocene subdivision, accepting physicochemical markers of GSSPs. This decision made the AWG's recent work on physicochemical markers for the Anthropocene more applicable.

Study sites for an Anthropocene candidate GSSP were known. Scientific evidence shows the details of the findings possible when studying physicochemical markers in fresh/soft sediments (Sirocko 2012; Gushulak et al. 2021). However, these studies are expensive, so the IUGS/ICS/AWG structure would find it challenging to pursue them. This difficulty diminished substantially when the HKW used funding obtained in 2018 (for its activities) to finance (via the AWG) a project to identify GSSP candidate sites for the base of the Anthropocene using physico-chemical markers. As a counterparty, the HKW got the right to exhibit the sediment cores holding these markers (Damianos 2021, p. 227ff.). Subsequently, the AWG would publish the findings in a special thematic issue of the journal, 'The Anthropocene Review', which began at the end of 2022.

2.7.2 *Narrative: Challenged Epistemic Leadership*

The IUGS, representing the global geological community and acting through its officers and constituent bodies, is deeply committed to the GTS as a recently agreed standardised scientific tool (Gibbard and Cohen 2008; Cohen et al. 2013). The most recent edition (Gradstein and Ogg 2020) is a compendium of nearly 1300 pages, involving about 100 senior authors and taking 8 years to compile. Seen from an IUGS perspective, the quest for a geological epoch Anthropocene came at an undue moment from an external promotor.

Before installing the AWG, the IUGS had to settle a fierce dispute about the fate of the Quaternary (Gibbard et al. 2005; Gibbard and Cohen 2008). Discussions followed about the base and structure of the Quaternary in the GTS (Head and Gibbard 2015; Gibbard and Head 2020) that occurred in the lifetime of the AWG. These disputes concerned various issues, like deleting the notion of Quaternary from the official geological nomenclature of the GTS and its replacement with 'late Neogene', moves of the IUGS to force the ICS to revise its ways of working[98] and the agreement on physico-chemical methodologies to establish the stages of the Holocene (Walker et al. 2018).

Having agreed in 2008 on the fate and general structure of the Quaternary, setting up a Working Group on the Anthropocene was suitable considering the usual pace of working groups. The former chair of the specific ICS commission dedicated to methodological questions of stratigraphy, Jan Zalasiewicz, guaranteed appropriate leadership regarding the expected methodological difficulties and his general interest in the subject.

The scientific details of the Quaternary/Neogene debate have limited bearing on the Anthropocene debate, but the agreement on physico-chemical methodologies in 2018 does. Also, the Anthropocene issue recalls issues of organisational discipline, peer consultation and agreed-upon procedures. The experience of the Quaternary/Neogene debate seemed particularly relevant for the incidences at the 35th IGC and the following Anthropocene debates. The ICS (seemingly) called the AWG 'to order' at the 35th IGC.

The experiences of the Quaternary/Neogene debate also seem relevant to cultural issues implied by geological topics. During the Quaternary/Neogene debate, it was argued that notions (e.g. Quaternary) should be employed that favour productive inter-disciplinary cooperation (Gibbard et al. 2005). This argumentation

[98] https://issc.uni-graz.at/Newsletters/ISSC_Newsletter_13.pdf (consulted 12 May 2024)

foreshadows similar arguments regarding the Anthropocene made in 2021 (Gibbard et al. 2021).

Having settled the Quaternary's structure, the circumstances of the Anthropocene debate stayed convoluted because the ICS/IUGS was discussing the Holocene subdivisions. Driven by recent technological advances, methodological issues regarding the GTS were debated in this context. In supporting the proposed Holocene subdivision, the IUGS agreed in 2018 to physico-chemical characteristics to determine a GSSP. This decision to adjust methodologies allowed the AWG after the 35th IGC to adapt its approach regarding how to mark the base of a geological Anthropocene.

In addition to these issues related to settling the stratigraphy of the Quaternary, the quest for a geological Anthropocene had further ramifications for the IUGS. In the first instance, modified methodologies were discussed by the AWG, e.g. using technofossils as primary markers within the GTS (Zalasiewicz et al. 2014; Zalasiewicz et al. 2017a, 2017b, 2017c). This positional argument was softened after the 35th IGC, favouring physico-chemical primary markers. In the second instance, the vocal advocacy of a geological Anthropocene came from a scientific community distant from Geology, the ESS. Given that this influential community was too close to Geology to be easily ignored by geologists, as is often the case for advocacy from Social Sciences or Humanities (Castree 2021), the IUGS/ICS community took up the challenge. In this manner, the IUGS/ICS community defended epistemic leadership for the GTS. Likely, seen from the IUGS/ICS perspectives, the AWG was 'rocking the boat' when stating some months before the 35th IGC, "The Anthropocene is functionally and stratigraphically distinct from the Holocene". (Waters et al. 2016 p. 1). The functional difference of Earth System dynamics is an ESS argument to which the AWG also stuck in its final proposal to the SQS (Waters et al. 2024b; Waters et al. 2024a). The IUGS quest for epistemic leadership also transpires from the phrasing of the public statement describing the decision to reject the AWG's proposal (IUGS 2024) p. 2):

> "An alternative narrative has therefore arisen in which the Anthropocene is not considered as a series/epoch (i.e., a chronostratigraphic unit and the corresponding geochronologic unit) but rather as an event, similar to the great transformative events in Earth history such as the Great Oxygenation (2.4-2.1 Ga), the Cambrian Explosion, or the Great Ordovician Biodiversification events. None of these major transformative events in Earth history are represented as chronostratigraphic units, and hence there has been no requirement for formal ratification. If so, the Anthropocene could be considered as an informal non-stratigraphical term"

2.7.3 Narrative: Evolution Pressures

Geology and ESS are distinct fields of knowledge within Earth Sciences, and the differences are multi-fold; for example, the prominent role of mathematically formulated theories and computer modelling in ESS. Also, the international and organisational cooperation networks are different and have limited overlap. For example, the World Meteorological Organisation (WMO) and global science programmes like IGBP are essential international fora for the ESS but less for the international geological community. Furthermore, consensus building in the ESS is shaped through the (intergovernmental) IPCC process—with high public visibility—compared to the traditional intra-disciplinary process of consecutive majorities of committees of peers for the IUGS/ICS. Finally, the IUGS has limited access to funding through intergovernmental arrangements, which, in turn, is commonplace when cooperating through structures like the WMO or the IGBP.

To the IUGS, its constituent bodies and communities, the idea of a geological epoch Anthropocene was like a 'cuckoo's egg'. It belonged to a different community[99] shaped by a distinct set of specific cooperation arrangements, material circumstances, institutions, societal practices and organisational cultures (Steffen et al. 2016). On the other hand, the IUGS's concerns for the methodological robustness of the GTS initially found little attention from an ESS perspective, starting with Crutzen's publication (Crutzen and Stoermer 2000; Crutzen 2002).

Beyond these irritations, which stemmed from the confrontation of distinct communities, parts of the geological communities (Autin and Holbrook 2012; Finney and Edwards 2016; De Wever and Finney 2018) seemingly misjudged the dynamics triggered by a geological notion that offers to integrate the descriptions of recent multi-faceted geological and socio-economical changes (Brauch 2021; Sklair 2021). The AWG had been partially receptive to this dynamic of post-normal science, as witnessed by its membership, publication activity, media activities and cooperation with third parties (e.g. HKW). The AWG may have misjudged the accumulated antagonism among IUGS/ICS officers. Subsequently, it was at least discomforting that the IUGS/ICS did not acknowledge the 2016 vote of the AWG. Instead, it summarised at the 35th IGC the dealings of the AWG for the professional public, saying: 'the preceding period being an establishment of the narrative concerning this possible division'[100] and finally insisted on its procedural leadership.

Considering these circumstances, the AWG adjusted its position in the 'economy of knowledge' to which it belongs, considering what a geological epoch Anthropocene may imply in terms of content, complying with established practices of the IUGS/ICS community and aggregating into the concept of a geological epoch Anthropocene recently agreed-upon stratigraphic methodology applied to Holocene subdivisions. In hindsight, it was the AWG's 'evolutionary response' configuring a specific 'synthetic thesis of truth':

- to aim at the mid-twentieth century as the time window to determine the geological base for the Anthropocene;
- to comply with the procedural rules of the IUGS/ICS community, including the assignment of voting rights;

[99] In the sense of 'economies of knowledge' (Renn 2020).

[100] https://stratigraphy.org/files/ICS-Business-IGC35.pdf

- to apply GSSP methodologies that had been recently accepted for the subdivision of the Holocene;
- to embrace the concept of the 'Great Acceleration' as the main feature, an Anthropocene GSSP should mark;
- to team up with a resourceful partner in the cultural/public sphere outside the geological community.

These choices reduced potential conflicts among geological communities, although they did not dispel them. The AWG fused various local economies of knowledge(s) of the Earth Sciences, emphasising, in the first instance, ESS and stratigraphy, as suggested by most AWG members some weeks before the 35th IGC (Steffen et al. 2016). The epistemic leadership of geological Stratigraphy to define the GTS was not questioned: geological Stratigraphy determines the geological concept of an Anthropocene. Finally, tying up the base of the Anthropocene with the historical period of the 'Great Acceleration' secures a persistent commitment of external communities, such as the HKW, representing Social Sciences and Humanities.

Nevertheless, this 'synthetic thesis of truth' did not move the majority views in SQS, ICS and IUGS and might have alienated them because of the remaining extra-geology features, including the use of nonconsolidated, fluffy lake sediments as location of a 'Golden Spike'. The verdict is net, paraphrasing the IUGS statement (IUGS 2024, p. 2), "major transformative events in Earth history do not require formal ratification as a chronostratigraphic unit", which is a very political view given humans are causing the major transformative event.

2.7.4 *Status of the AWG*

Throughout its lifespan, 2009–2023, the AWG was a peculiar chimaera. On one side, it was a body of academic peers dedicated to pursuing a methodological issue in their disciplines. On the other hand, the AWG was tied to the bureaucratic processes of the IUGS/ICS system, overseeing a specific methodological question with interdisciplinary emphasis. As a bureaucratic body, the AWG ceased existence. However, the academic peer group can continue to function.

When finalising this chapter, the ICS/IUGS management sealed the fate of AWG's formal proposal (Witze 2024b). How science-policy dynamics evolve around the Anthropocene concept in consecutive years is speculative, including in which form the AWG might continue its work. The baseline is relatively straightforward. The ICS/IUGS management promotes an Anthropocene event (compromising on the misnomer '-cene'). It eclipses the insight that the Earth System's dynamics currently experience a state shift, a paramount feature when taking an ESS perspective (Rockström et al. 2024). The AWG's substantial scientific legacy supports this ESS perspective, e.g. (Zalasiewicz et al. 2019; Waters et al. 2016; 2023, 2024a, 2024b). A further essential part of the AWG legacy is the established cooperation beyond Earth Science communities, raising awareness of the societal relevance of Earth Sciences expertise for a globalised economy, culture and stewardship for the Earth System. Happily, the advocates of the event notion, e.g. (Maslin and Lewis 2015; Koster et al. 2024), share these perspectives.

The AWG showed how to navigate a multifaceted global environment by transgressing the borders of traditional (geological) sciences. In this context, the AWG featured intellectual resilience and learning, the ability to enter strategic coalitions and the mobilisation of external support. The dealings of the AWG show path-depending triangulation of imposed constraints, external opportunities and material contexts. The AWG's findings evolve by combining scientific insights aggregated around its (bureaucratic) mandate, which illustrates how Mario Bunge's 'synthetic thesis of truth' (Marone et al. 2019) applies to knowledge building in geosciences.

The AWG is a specific economy of knowledge in the sense that J. Renn's theory (Renn 2020) describes the evolution of knowledge. The AWG had a substantial inter-disciplinary and public impact, which likely caused the AWG's limited intra-disciplinary influence. Nevertheless, the AWG helped to modernise thinking of geological communities; in the wording of the final tune of IUGS's decision (IUGS 2024, p. 3), "the Anthropocene as a concept will continue to be widely used not only by Earth and environmental scientists but also by social scientists, politicians and economists, as well as by the public at large. As such, it will remain an invaluable descriptor in human-environment interactions". Considering the likely societal impact of ongoing planetary-scale anthropogenic change (Kotz et al. 2024), the IUGS statement is somewhat encouraging because it marks a substantial step for the hegemonic thinking of this INGO. However, it falls critically short of societal needs (Barnosky and Hannibal 2024).

Nonetheless, given favourable external circumstances, such as the early and lasting cooperation with the HKW, the AWG offers a significant example of how the globalisation of science happens. This achievement contrasts favourably with "pathological path dependency in institutions, practices and ideas" (Dryzek and Pickering 2018, p. 151) of conventional Holocene milieus, and of which the ICS/IUGS system is an example—belated start to handle the Anthropocene question, using established practices, sticking to conventional solutions and missing the essential feature (Zalasiewicz 2024). One may wonder what an AWG could deliver if set up *sui generis* by the International Science Council.

References

Asafu-Adjaye, John, Linus Blomqvist, Stewart Brand, Barry Brook, Ruth Defries, Erle C. Ellis, David Keith, et al. 2015. *An Ecomodernist manifesto*, 31. Oakland: Breakthrough Institute. https://www.ecomodernism.org/manifesto-english (10th March 2016)

Autin, Whitney J., and John M. Holbrook. 2012. Is the Anthropocene an issue of stratigraphy or pop culture? *GSA Today* 22:60–61. https://doi.org/10.1130/G153GW.1.

Baker, Victor R. 1999. Geosemiosis. *Geological Society of America Bulletin* 111:633–645. https://doi.org/10.1130/0016-7606(1999)111<0633:G>2.3.CO;2.

Baldwin, Andrew, and Bruce Erickson. 2020. Introduction: Whiteness, coloniality, and the Anthropocene. *Environment and Planning D: Society and Space* 38:3–11. https://doi.org/10.1177/0263775820904485.

Ballantyne, Nathan. 2019. Epistemic trespassing. *Mind* 128:367–395. https://doi.org/10.1093/mind/fzx042.

Barnosky, Anthony D., and Ellen Hannibal. 2024. Despite official vote, the evidence of the Anthropocene is clear. *Yale Environment* 360. https://e360.yale.edu/features/anthropocene-denied (accessed 2th May 2024)

Barton, Jeni. 2019. The geo-politics of the anthropocene: Using stratigraphy to naturalize the anthropocene as a formal geological unit. *Analecta Hermeneutica* 10:17–31.

Bauer, Andrew M., Matthew Edgeworth, Lucy E. Edwards, Erle C. Ellis, Philip L. Gibbard, and Dorothy J. Merritts. 2021. Anthropocene: Event or epoch? *Nature* 597:332–332. https://doi.org/10.1038/d41586-021-02448-z.

Benner, Susanne, Gregor Lax, Paul J. Crutzen, Ulrich Pöschl, Jos Lelieveld, and Hans Günter Brauch. 2021. *Paul J. Crutzen and the Anthropocene: A new epoch in Earth's history*. Susanne Benner, Gregor Lax, Paul J. Crutzen, Ulrich Pöschl, Jos Lelieveld, and Hans Günter Brauch. Vol. 1. The Anthropocene: Politik—Economics—Society—Science. Cham: Springer International Publishing. doi:https://doi.org/10.1007/978-3-030-82202-6.

Bohle, Martin. 2023. Crafting attributes of a geological now. *Earth Science, Systems and Society* 3:1–13. https://doi.org/10.3389/esss.2023.10089.

Bowen, D. Q. 1999. A revised correlation of quaternary deposits in the British Isles. In *A revised correlation of quaternary deposits in the British Isles*, ed. D. Q. Bowen. London, UK: Geological Society of London. https://doi.org/10.1144/SR23.

Brauch, Hans Günter. 2021. The anthropocene concept in the natural and social sciences, the humanities and law—A bibliometric analysis and a qualitative interpretation (2000–2020). In *Paul Crutzen and the Anthropocene; a new epoch in Earth's history*, 289–438. Cham (CH): Springer International Publishing. https://doi.org/10.1007/978-3-030-82202-6_22.

Castree, Noel. 2021. Framing, deframing and reframing the Anthropocene. *Ambio*. Springer Netherlands 50:1788. https://doi.org/10.1007/s13280-020-01437-2.

Cohen, K. M., S. C. Finney, Philip L. Gibbard, and J.-X. Fan. 2013. The ICS international chronostratigraphic chart. *Episodes* 36:199–204. https://doi.org/10.18814/epiiugs/2013/v36i3/002.

Crutzen, Paul J., and Eugen F. Stoermer. 2000. The "Anthropocene". *The Global Change Newsletter* 41:17–18.

Crutzen, Paul J. 2002. Geology of mankind. Nature. Vol 415:23

Damianos, Alexander. 2021. *Ratifying the Anthropocene: A study of the Anthropocene working Group's ongoing effort to formalize the Anthropocene as a unit of the geologic time scale*. The London School of Economics and Political Science. 304p. https://etheses.lse.ac.uk/4351/ (accessed 28th Sep. 2022)

Denevan, William M. 2011. The "pristine myth" revisited. *Geographical Review* 101:576–591. https://doi.org/10.1111/j.1931-0846.2011.00118.x.

Dodick J, Orion N (2003) Geology as an historical science: Its Perception within Science and the Education System. Sci Educ 12:197–211. https://doi.org/10.1023/A:1023096001250

Dryzek, John S., and Jonathan Pickering. 2018. *The politics of the Anthropocene*. Oxford: Oxford University Press. https://doi.org/10.1093/oso/9780198809616.001.0001.

Edgeworth, Matt, Erle C. Ellis, Philip L. Gibbard, Cath Neal, and Michael Ellis. 2019. The chronostratigraphic method is unsuitable for determining the start of the Anthropocene. *Progress in Physical Geography: Earth and Environment* 43:334–344. https://doi.org/10.1177/0309133319831673.

Edgeworth, Matthew, Philip L. Gibbard, Michael Walker, Dorothy Merritts, Stanley Finney, and Mark Maslin. 2023. The stratigraphic basis of the anthropocene event. *Quaternary Science Advances* 11. Elsevier Ltd:100088. https://doi.org/10.1016/j.qsa.2023.100088.

Ellenberger F (1978) The first international geological congress - Paris, 1878. Episodes 1:20–24. https://doi.org/10.18814/epiiugs/1978/v1i2/004

Ellis, Erle C. 2024. The Anthropocene condition: Evolving through social–ecological transformations. *Philosophical Transactions of the Royal Society B: Biological Sciences* 379:20220255. https://doi.org/10.1098/rstb.2022.0255.

Finney, Stanley C. 2014. The ‘Anthropocene’ as a ratified unit in the ICS international chronostratigraphic chart: Fundamental issues that must be addressed by the task group. *Geological Society, London, Special Publications* 395:23–28. https://doi.org/10.1144/SP395.9.

Finney, Stanley C., and Lucy E. Edwards. 2016. The “Anthropocene” epoch: Scientific decision or political statement? *GSA Today* 26:4–10. https://doi.org/10.1130/GSATG270A.1.

Finney, Stanley C., and Asier Hilario. 2018. *GSSPs as international geostandards and as global geoheritage*. Geoheritage: Assessment, Protection, and Management. Elsevier Inc. https://doi.org/10.1016/B978-0-12-809531-7.00010-1.

Folke, Carl, and Johan Rockström. 2011. 3rd Nobel laureate symposium on global sustainability: Transforming the world in an era of global change. *Ambio* 40:717–718. https://doi.org/10.1007/s13280-011-0190-0.

Foster J (2018) Let’s not talk about the anthropocene. Analecta Hermeneutica 10:1–37

Frodeman, Robert. 1995. Geological reasoning: Geology as an interpretive and historical science. *Geological Society of America Bulletin.* 107:960. https://doi.org/10.1130/0016-7606(1995)107<0960:grgaai>2.3.co;2.

Funtowicz, Silvio, and Jerome Ravetz. 2003. Post-normal science. In *Companion to environmental studies*, 443–447. Routledge. https://doi.org/10.4324/9781315640051-89.

Gibbard, Philip L., Andrew M. Bauer, Matthew Edgeworth, William F. Ruddiman, Jacquelyn L. Gill, Dorothy J. Merritts, Stanley C. Finney, et al. 2021. A practical solution: The Anthropocene is a geological event, not a formal epoch. *Episodes* 1964:1–9. https://doi.org/10.18814/epiiugs/2021/021029.

Gibbard, Philip L., and Kim M. Cohen. 2008. Global chronostratigraphical correlation table for the last 2.7 million years. *Episodes* 31:243–247. https://doi.org/10.18814/epiiugs/2008/v31i2/011.

Gibbard, Philip L., and M. J. Head. 2020. The quaternary period. In *Geologic Time Scale 2020*, 1217–1255. Elsevier. https://doi.org/10.1016/B978-0-12-824360-2.00030-9.

Gibbard, Philip L., Alan G. Smith, Jan A. Zalasiewicz, Tiffany L. Barry, David Cantrill, Angela L. Coe, John C. W. Cope, et al. 2005. What status for the quaternary? *Boreas* 34:1–6. https://doi.org/10.1080/03009480510012854.

Gibbard, Philip L., and M. J. C. Walker. 2014. The term ‘Anthropocene’ in the context of formal geological classification. *Geological Society, London, Special Publications* 395:29–37. https://doi.org/10.1144/SP395.1.

Gradstein, F. M., and J. G. Ogg. 2020. The chronostratigraphic scale. In *Geologic Time Scale 2020*, 21–32. Elsevier. https://doi.org/10.1016/B978-0-12-824360-2.00002-4.

Gushulak, Cale A. C., Matthew Marshall, Brian F. Cumming, R. Brendan Llew-Williams, Timothy Patterson, and Francine M. G. McCarthy. 2021. *Siliceous algae response to the “great acceleration” of the mid-20th century in Crawford Lake*. Ontario, Canada: A potential candidate for the Anthropocene GSSP. Anthropocene Review. https://doi.org/10.1177/20530196211046036.

Hall, D. H. 1976. *History of the earth sciences during the scientific and industrial revolutions with special emphasis on the physical geosciences*. Amsterdam: Elsevier Scientific Publishing Company.

Hazen, Robert M. 2012. *The story of earth: The first 4.5 billion years, from stardust to living planet*. Viking Penguin Grroup.

Head, Martin J., and Philip L. Gibbard. 2015. Formal subdivision of the quaternary system/period: Past, present, and future. *Quaternary International* 383. Elsevier Ltd:4–35. https://doi.org/10.1016/j.quaint.2015.06.039.

Head, Martin J., Philip L. Gibbard, and Thijs van Kolfschoten. 2013. Quaternary system/period: Current status and future challenges. *Ciencias da Terra* 18 (18): 77–80.

Head, Martin J., Jan A. Zalasiewicz, Colin N. Waters, Simon D. Turner, Mark Williams, Anthony D. Barnosky, Will Steffen, et al. 2022. The Anthropocene is a prospective epoch/series, not a geological event. *Episodes* 46:229. https://doi.org/10.18814/epiiugs/2022/022025.

Hughes, Lesley, and Martin Rice. 2023. Will Steffen (1947–2023). *Nature* 615:29–29. https://doi.org/10.1038/d41586-023-00519-x.

IPCC. 2018. *Global warming of 1.5°C*. Cambridge University Press. https://doi.org/10.1017/9781009157940.

Ismail-Zadeh, Alik. 2016. Geoscience international: The role of scientific unions. *History of Geo- and Space Sciences* 7:103–123. https://doi.org/10.5194/hgss-7-103-2016.

IUGS. 2024. The Anthropocene. https://www.iugs.org/_files/ugd/f1fc07_40d1a7ed58de458c9f8f24de5e739663.pdf (accessed 23rd March 2024)

De Jouvancourt, Pierre. 2024. Qui a tué l'Anthropocène? *Terrestres*. https://www.terrestres.org/2024/04/08/qui-a-tue-lanthropocene/ (accessed 1st May 2024)

Koster, Emlyn. 2022. Public-minded reflections from the Anthropocene working group meeting in Germany. *Episodes* 46:317. https://doi.org/10.18814/epiiugs/2022/022033.

Koster, Emlyn, and Philip L. Gibbard. 2024. The most consequential ethical decision for geoscience. EGU General Assembly 2024, Vienna, Austria, 14–19 Apr 2024. https://meetingorganizer.copernicus.org/EGU24/EGU24-12918.html (accessed 15th April 2024)

Koster, Emlyn, Philip L. Gibbard, and Mark Maslin. 2023. Optimising the Anthropocene definition: An epistemological view with briefings on four 2022-23 conferences. *Episodes* 46:1–12. https://doi.org/10.18814/epiiugs/2023/023005.

———. 2024. The Anthropocene event as a cultural zeitgeist in the earth-human ecosystem. *Journal of Geoethics and Social Geosciences* 01:1–41. https://doi.org/10.13127/jgsg-43.

Kotz, Maximilian, Anders Levermann, and Leonie Wenz. 2024. The economic commitment of climate change. *Nature* 628. Springer US:551–557. https://doi.org/10.1038/s41586-024-07219-0.

Langmuir, Charles, and Wally Broecker. 2012. *How to build a habitable planet?* Princeton: Princeton University Press.

Lewis, Simon L., and Mark A. Maslin. 2015. A transparent framework for defining the Anthropocene epoch. *The Anthropocene Review* 2:128–146. https://doi.org/10.1177/2053019615588792.

Lundershausen, Johannes. 2018a. *Engaging Anthropocene science: Perspectives on the role of geoscientific practices on Anthropocene debates*. 134p. Tübingen.

———. 2018b. Marking the boundaries of stratigraphy: Is stratigraphy able and willing to define, describe and explain the Anthropocene? *Geo: Geography and Environment* 5:e00055. https://doi.org/10.1002/geo2.55.

———. 2018c. The Anthropocene working group and its (inter-)disciplinarity. Sustainability: Science. *Practice and Policy* 14:31–45. https://doi.org/10.1080/15487733.2018.1541682.

Marone, Luis, Javier Lopez de Casenave, and Rafael González del Solar. 2019. *The synthetic thesis of truth helps mitigate the "reproducibility crisis" and is an inspiration for predictive ecology, Humanities Journal of Valparaiso*. University of Valparaiso363–376. https://doi.org/10.22370/rhv2019iss14pp363-376.

Maslin, Mark A., and Simon L. Lewis. 2015. Anthropocene: Earth system, geological, philosophical and political paradigm shifts. *The Anthropocene Review* 2:108–116. https://doi.org/10.1177/2053019615588791.

Mogk, David W. 2020. The Intersection of Geoethics and Diversity in the Geosciences. In *Geological Society*, SP508–2020–66. London: Special Publications. https://doi.org/10.1144/SP508-2020-66.

Molina, Mario J., and F. Sherwood Rowland. 2010. Anthropocene man. Nature 467. Nature publishing group, a division of Macmillan Publishers Limited. *All Rights Reserved* 467:S10. https://doi.org/10.1038/467S10a.

Laureates, Nobel. 2011. The Stockholm memorandum. *Ambio* 40:781–785. https://doi.org/10.1007/s13280-011-0187-8.

Phillips, Jonathan. 2012. Storytelling in earth sciences: The eight basic plots. Earth-Science Reviews 115. Elsevier B.V.: 153–162. https://doi.org/10.1016/j.earscirev.2012.09.005.

Price, Simon J., Jonathan R. Ford, Anthony H. Cooper, and Catherine Neal. 2011. Humans as major geological and geomorphological agents in the Anthropocene: The significance of artificial ground in Great Britain. *Philosophical Transactions of the Royal Society A: Mathematical, Physical and Engineering Sciences* 369:1056–1084. https://doi.org/10.1098/rsta.2010.0296.

Renn, Jürgen. 2020. *The evolution of knowledge–rethinking science for the Anthropocene*, 561. Oxford, UK: Princeton University Press.

Rockström, Johan, Louis Kotzé, Svetlana Milutinović, Frank Biermann, Victor Brovkin, Jonathan Donges, Jonas Ebbesson, et al. 2024. The planetary commons: A new paradigm for safeguarding earth-regulating systems in the Anthropocene. *Proceedings of the National Academy of Sciences* 121:1–10. https://doi.org/10.1073/pnas.2301531121.

Rosol, Christoph, Georg N. Schäfer, Simon D. Turner, Colin N. Waters, Martin J. Head, Jan Zalasiewicz, Carlina Rossée, Jürgen Renn, Katrin Klingan, and Bernd M. Scherer. 2023. Evidence and experiment: Curating contexts of Anthropocene geology. *The Anthropocene Review* 10:330–339. https://doi.org/10.1177/20530196231165621.

Ruddiman, William F. 2019. Reply to Anthropocene working group responses. *Progress in Physical Geography: Earth and Environment* 030913331983992:345. https://doi.org/10.1177/0309133319839926.

Rull, Valentí. 2018. What if the 'Anthropocene' is not formalized as a new geological series/epoch? *Quaternary* 1:24. https://doi.org/10.3390/quat1030024.

Schofer, Evan. 2003. The Global Institutionalization of Geological Science, 1800 to 1990. *American Sociological Review* 68:730–759. https://doi.org/10.2307/1519760.

Seddon, G. 1996. Thinking like a geologist: The culture of geology. *Australian Journal of Earth Sciences* 43:487.

Sirocko, Frank. 2012. Wetter, Klima. Menschheitentwicklung, 208. Theiss.

Sklair, Leslie. 2021. In *The Anthropocene in global media*, ed. Leslie Sklair. Abingdon, Oxon: Routledge. https://doi.org/10.4324/9780429355202.

Steffen, Will, Wendy Broadgate, Lisa Deutsch, Owen Gaffney, and Cornelia Ludwig. 2015. The trajectory of the Anthropocene: The great acceleration. *The Anthropocene Review* 2:81–98. https://doi.org/10.1177/2053019614564785.

Steffen, Will, Reinhold Leinfelder, Jan Zalasiewicz, Colin N. Waters, Mark Williams, Colin Summerhayes, Anthony D. Barnosky, et al. 2016. Stratigraphic and earth system approaches to defining the Anthropocene. *Earth's Future* 4:324–345. https://doi.org/10.1002/2016EF000379.

Steffen, Will, Katherine Richardson, Johan Rockström, Hans Joachim Schellnhuber, Opha Pauline Dube, Sébastien Dutreuil, Timothy M. Lenton, and Jane Lubchenco. 2020. The emergence and evolution of earth system science. *Nature Reviews Earth & Environment* 1:54–63. https://doi.org/10.1038/s43017-019-0005-6.

Thomas, Julia Adeney. 2022. In *Altered earth*, ed. Julia Adeney Thomas. Cambridge, UK: Cambridge University Press. https://doi.org/10.1017/9781009042369.

Trischler, Helmuth. 2016. The Anthropocene. *NTM Zeitschrift für Geschichte der Wissenschaften, Technik und Medizin* 24:309–335. https://doi.org/10.1007/s00048-016-0146-3.

Turner, Simon D., Collins Waters, Jan Zalasiewicz, and MartinJ. Head. 2024. *What the Anthropocene's critics overlook- and why it really should be a new geological epoch*. The Conversation. https://theconversation.com/what-the-anthropocenes-critics-overlook-and-why-it-really-should-be-a-new-geological-epoch-225493 (accessed 14th March 2024)

Walker, Mike, Martin J. Head, Max Berkelhammer, Svante Björck, Hai Cheng, Les Cwynar, David Fisher, et al. 2018. Formal ratification of the subdivision of the Holocene series/epoch (quaternary system/period): Two new global boundary Stratotype sections and points (GSSPs) and three new stages/subseries. *Episodes* 41:213–223. https://doi.org/10.18814/epiiugs/2018/018016.

Wallenhorst, Nathanaël. 2023. *A critical theory for the Anthropocene. Anthropocene–Humanities and Social Sciences*. Cham: Springer International Publishing. https://doi.org/10.1007/978-3-031-37738-9.

Warde, Paul, Libby Robin, and Sverker Sörlin. 2017. Stratigraphy for the Renaissance: Questions of expertise for 'the environment' and 'the Anthropocene. *The Anthropocene Review* 4:246–258. https://doi.org/10.1177/2053019617738803.

Waters, Colin N., Simon D. Turner, Jan Zalasiewicz, and Martin J. Head. 2023. Candidate sites and other reference sections for the global boundary Stratotype section and point of the Anthropocene series. *The Anthropocene Review* 205301962211364:3. https://doi.org/10.1177/20530196221136422.

Waters, Colin N., Simon Turner, An Zhisheng, Anthony Barnosky, Alejandro Cearreta, Andrew Cundy, Ian Fairchild, et al. 2024a. The Anthropocene Epoch and Crawfordian Age: proposals by the Anthropocene Working Group Part 2: Descriptions of the proposed Crawford Lake GSSP and supporting SABSs. https://eartharxiv.org/repository/view/6963/ (accessed 17th April 2024)

———. 2024b. The Anthropocene Epoch and Crawfordian Age: proposals by the Anthropocene Working Group Part 1: Anthropocene Series/Epoch: stratigraphic context and justification of rank. https://eartharxiv.org/repository/view/6954/ (accessed 17h April 2024)

Waters, Colin N., Mark Williams, Jan Zalasiewicz, Simon D. Turner, Anthony D. Barnosky, Martin J. Head, Scott L. Wing, et al. 2022. Epochs, events and episodes: Marking the geological impact of humans. *Earth-Science Reviews* 234:104171. https://doi.org/10.1016/j.earscirev.2022.104171.

Waters, Colin N., Jan Zalasiewicz, Colin Summerhayes, Anthony D. Barnosky, A. Clément Poirier, Alejandro Cearreta Gauszka, et al. 2016. The Anthropocene is functionally and stratigraphically distinct from the Holocene. *Science* 351:aad2622–aad2622. https://doi.org/10.1126/science.aad2622.

Wenninger, Andreas, Fabienne Will, Sascha Dickel, Sabine Maasen, and Helmuth Trischler. 2019. 1. Ein- und Ausschließen: Evidenzpraktiken in der Anthropozändebatte und der Citizen Science. In *Wissen und Begründen*, 31–58. Nomos Verlagsgesellschaft mbH & Co. KG. https://doi.org/10.5771/9783748903383-31.

De Wever, Patrick, and Stan Finney. 2018. The anthropocene: A geological or societal subject? *Biodiversity and Evolution. Elsevier* 251–264. https://doi.org/10.1016/B978-1-78548-277-9.50014-0.

Will, Fabienne. 2021. Evidenz für das Anthropozän -Wissensbildung und Aushandlungsprozesse an der Schnittstelle von Natur-, Geistes- und Sozialwissenschaften. 354p. Göttingen: Vandenhoeck & Ruprecht.

Witze, Alexandra. 2023. This quiet lake could mark the start of a new Anthropocene epoch. *Nature* 619:441–442. https://doi.org/10.1038/d41586-023-02234-z.

———. 2024a. Geologists reject the Anthropocene as Earth's new epoch — After 15 years of debate. *Nature* 627:249. https://doi.org/10.1038/d41586-024-00675-8.

———. 2024b. It's final: The Anthropocene is not an epoch, despite protest over vote. *Nature*. https://doi.org/10.1038/d41586-024-00868-1.

Zalasiewicz, Jan. 2008. *The earth after us: What legacy will humans leave in the rocks?* Oxford: Oxford University Press.

Zalasiewicz, Jan. 2022. Science: Old and new patterns of the Anthropocene. In *Altered earth*, ed. Julia Adeney Thomas, 21–50. Cambridge (UK): Cambridge University Press. https://doi.org/10.1017/9781009042369.004.

Zalasiewicz, Jan. 2024. Why criticisms of the proposed Anthropocene epoch miss the point. *New Scientist*. https://www.newscientist.com/article/mg26234900-100-why-criticisms-of-the-proposed-anthropocene-epoch-miss-the-point/ (accessed 11th May 2024)

Zalasiewicz, Jan, Colin N. Waters, Erle C. Ellis, Martin J. Head, Davor Vidas, Will Steffen, Julia Adeney Thomas, et al. 2021. The Anthropocene: Comparing its meaning in geology (Chronostratigraphy) with conceptual approaches arising in other disciplines. *Earth's Future* 9. https://doi.org/10.1029/2020EF001896.

Zalasiewicz, Jan, Colin N. Waters, Colin P. Summerhayes, Alexander P. Wolfe, Anthony D. Barnosky, Alejandro Cearreta, Paul Crutzen, et al. 2017a. The Working Group on the Anthropocene: Summary of evidence and interim recommendations. Anthropocene 19. Elsevier B.V.: 55–60. https://doi.org/10.1016/j.ancene.2017.09.001.

Zalasiewicz, Jan, Colin N. Waters, Mark Williams, Anthony D. Barnosky, Alejandro Cearreta, Paul Crutzen, Erle Ellis, et al. 2015. When did the Anthropocene begin? A mid-twentieth century boundary level is stratigraphically optimal. *Quaternary International* 383:196–203. https://doi.org/10.1016/j.quaint.2014.11.045.

Zalasiewicz, Jan, Colin N. Waters, Mark Williams, and Colin Summerhayes. 2019. In *The Anthropocene as a geological time unit*, ed. Jan Zalasiewicz, Colin N. Waters, Mark Williams, and Colin P. Summerhayes. Cambridge University Press. https://doi.org/10.1017/9781108621359.

Zalasiewicz, Jan, Colin Waters, Simon Turner, Mark Williams, and Martin J. Head. 2023. Anthropocene working group. In *Handbook of the Anthropocene*, 315–321. Cham: Springer International Publishing. https://doi.org/10.1007/978-3-031-25910-4_48.

Zalasiewicz, Jan, Collins Waters, Mark Williams, Collin Summerhayes, and Philip L. Gibbard. 2017b. The geological Anthropocene born in Burlington house. Geoscientist online. https://www.geolsoc.org.uk/Geoscientist/Archive/December-January-2017-18/The-Geological-Anthropocene (accessed 5th Feb. 2022)

Zalasiewicz, Jan, and Mark Williams. 2012. The goldilocks planet. *Oxford University Press.* https://doi.org/10.1093/oso/9780199593576.001.0001.

Zalasiewicz, Jan, Mark Williams, Alan Haywood, and Michael Ellis. 2011. The Anthropocene: A new epoch of geological time? *Philosophical Transactions of the Royal Society A: Mathematical, Physical and Engineering Sciences* 369:835–841. https://doi.org/10.1098/rsta.2010.0339.

Zalasiewicz, Jan, Mark Williams, Alan Smith, Tiffany L. Barry, Angela L. Coe, Paul R. Bown, Patrick Brenchley, et al. 2008. Are we now living in the Anthropocene. *GSA Today* 18:4. https://doi.org/10.1130/GSAT01802A.1.

Zalasiewicz, Jan, Mark Williams, Colin N. Waters, Anthony D. Barnosky, and Peter Haff. 2014. The technofossil record of humans. *The Anthropocene Review* 1:34–43. https://doi.org/10.1177/2053019613514953.

Zalasiewicz, Jan, Mark Williams, Colin N. Waters, Anthony D. Barnosky, John Palmesino, Ann-Sofi Rönnskog, Matt Edgeworth, et al. 2017c. Scale and diversity of the physical technosphere: A geological perspective. *The Anthropocene Review* 4:9–22. https://doi.org/10.1177/2053019616677743.

Zen, E-An. 1993. The citizen-geologist. *GSA Today* 3:2–3.

Chapter 3
The AWG and the Scientific Debate About the Anthropocene

3.1 Introduction

The question of when the Anthropocene began and whether it should be considered a distinct chronostratigraphic epoch has been addressed not only by geologists but also by scientists and scholars from other fields, such as Earth Systems science and environmental studies.[1] This has resulted in a wide range of interpretations: Some, following the initial suggestion by Stoermer and Crutzen (2000), argue that the Anthropocene began with the Industrial Revolution, while others propose an earlier date, such as the onset of agriculture (Ruddiman 2003) or even the emergence of human civilization; however, there are also arguments that associate it with the "Great Acceleration" of the twentieth century (McNeill and Engelke 2016; Steffen et al. 2007). Since the year 2009, the *Anthropocene Working Group* (AWG) of the *Subcommission on Quaternary Stratigraphy* (SQS) in the *International Union of Geological Sciences* (IUGS) has shaped the Anthropocene debate from a geological but also from a more interdisciplinary perspective.

While the preceding chapter has focused on the role of the AWG within the institutional context, this chapter investigates how the AWG members have addressed the Anthropocene in their publications, whether and how they have thereby established a scientific community and to what extent their works have influenced and shaped the more general scientific discourse about the Anthropocene. Comprising a wide range of disciplines with a focus on geoscience, the AWG provides a large spectrum of expertise for the analysis of anthropogenic planetary change. However,

---

This chapter is authored by Boris Holzer.

---

[1] I want to thank my coauthors of this volume, Martin Bohle, Leslie Sklair, and Fabienne Will, for their valuable input and suggestions. I am also grateful to Sophia Luhr, Esra Öndüc, and Max Wagner for their research assistance.

M. Bohle et al., *The Anthropocene Working Group and the Global Debate Around a New Geological Epoch*, Emerging Globalities and Civilizational Perspectives, https://doi.org/10.1007/978-3-031-85175-9_3

it operates under a mandate from the ICS and, thus, primarily within the disciplinary framework of the geological sciences. The decision to frame the Anthropocene question in terms of the Geological Time Scale meant that stratigraphic methods and the demarcation of chronostratigraphic units became a focus of the AWG, but this does not exhaust the research fields of AWG members, who have published scientific contributions to the ongoing Anthropocene debate and other research fields, both jointly and individually. This chapter takes stock of these publications of AWG members and analyzes them regarding the forms of collaboration that underpin them. This chapter will not evaluate the merit of individual arguments and claims but reconstruct how the Anthropocene was established as a topic of scientific debate for which the AWG served as an institutional focus. From a microsociological perspective, science is usually conceived as a practice that relies on the cooperation of scientists, which may occur in situations of physical co-presence, such as the laboratory (Latour and Woolgar 1979; Knorr-Cetina 1981; Latour 1983). However, scientific cooperation may also occur at a distance, either through written communication in the tradition of the "Republic of Letters" (van Vugt 2022) or through various forms of remote communication when scientists collaborate to jointly publish a research statement based on their individual contributions. From a macrosociological perspective, such publications are the basic elements of science as a subsystem of society (Luhmann 1992). They form a cohesive but temporalized system of communication based on citations: Publications refer to other publications and, thereby, weave a network of interconnections (Leydesdorff 1998; Stichweh 2014).

In the following, I will first survey the disciplinary and institutional affiliations of current and former AWG members as well as indicators of their reputation in their research areas. Second, I will analyze the patterns of collaboration and citation in a bibliographic corpus of AWG members from 2000 to 2020 drawn from the *Web of Science* database. Third, this collection of publications will be used to retrace the thematic evolution of the AWG's contribution to the Anthropocene debate. Fourth and finally, I will relate these findings of an "in-group" corpus of publications to another "out-group" collection of publications mentioning the Anthropocene (also from *Web of Science*) to examine how visible the AWG members' publications are in the broader scientific context and to what degree the AWG has been able to shape it, for instance, by influential publications. In the combination of these different perspectives, this chapter aims to shed light on the intra- and cross-disciplinary resonance of the Anthropocene and the role of the AWG in establishing the Anthropocene as a research theme in the geosciences and beyond.

3.2 Disciplines, Regions, and Reputation

Despite its interdisciplinary outlook, it is hardly surprising that the AWG should primarily include members from the disciplines of geology and the geosciences more broadly. The ongoing debate about what kind of change and "marker" should

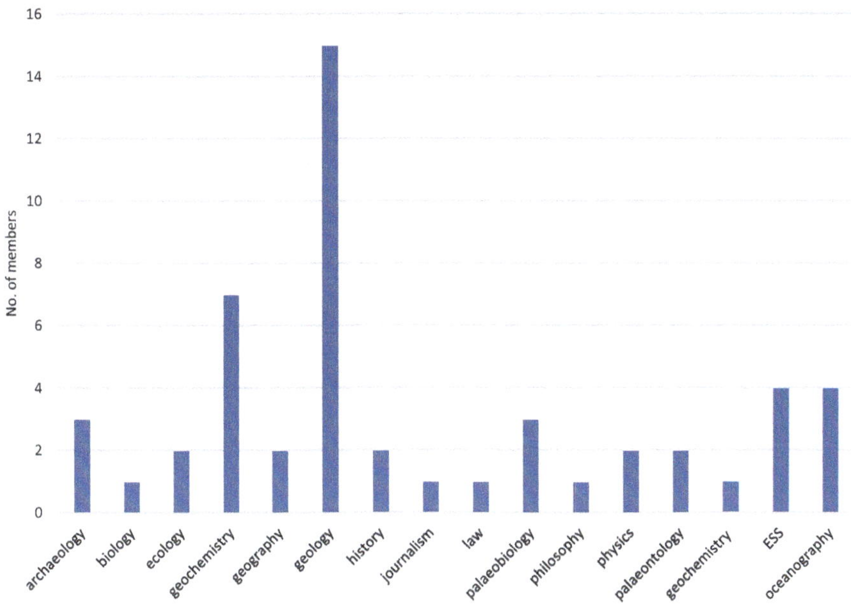

Fig. 3.1 Disciplines represented in the AWG, 2009–2023

define the onset of the Anthropocene is reflected in the participation of archaeologists, biologists, climate and atmospheric scientists, and oceanographers (Fig. 3.1). The humanities are also represented by individual scholars from history, law, and philosophy, but they are clearly in a more advisory role.[2]

A comparison of snapshots between the years 2009 and 2023 (Fig. 3.2) shows that over the years, the disciplinary matrix of the group has slightly expanded through the recruitment of new members, while its focus on the geosciences has been affirmed.

The geographical distribution of the members' country affiliations during the period 2009–2023 shows that most members are located in the Western hemisphere (Fig. 3.3). More than half of them work at universities in the United Kingdom and the USA, about a fifth at European institutions. Asian and Latin American scholars are a small minority.[3]

This distribution reflects but does not exactly mirror the geographical distribution of the international (geo-)scientific community. Its particular bias has much to do with the fact that the AWG grew out of a small group of founding members who had to recruit new members based on their research profiles but also with an eye toward how they could fit into and complement the existing group. Such a recruitment process often relies on existing personal networks and collaborations and,

[2] See also Chaps. 2 and 4 for the institutional and scientific environments of the AWG.

[3] See SQS (2024) for a current list of AWG members.

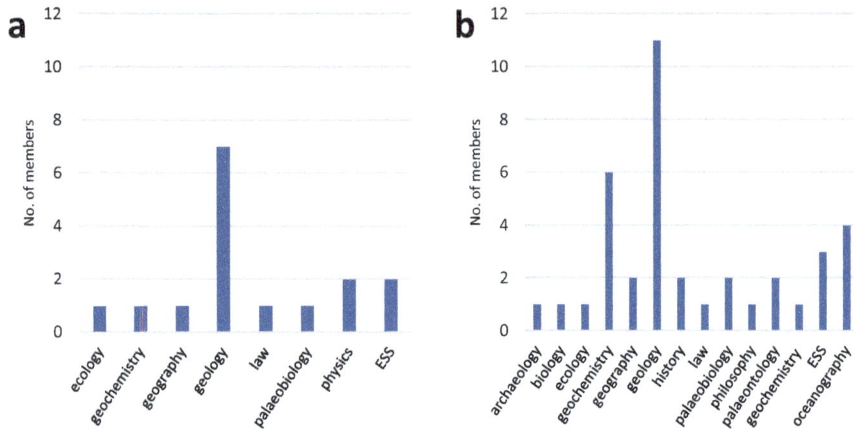

Fig. 3.2 Disciplines of AWG members, (**a**) 2009 and (**b**) 2023

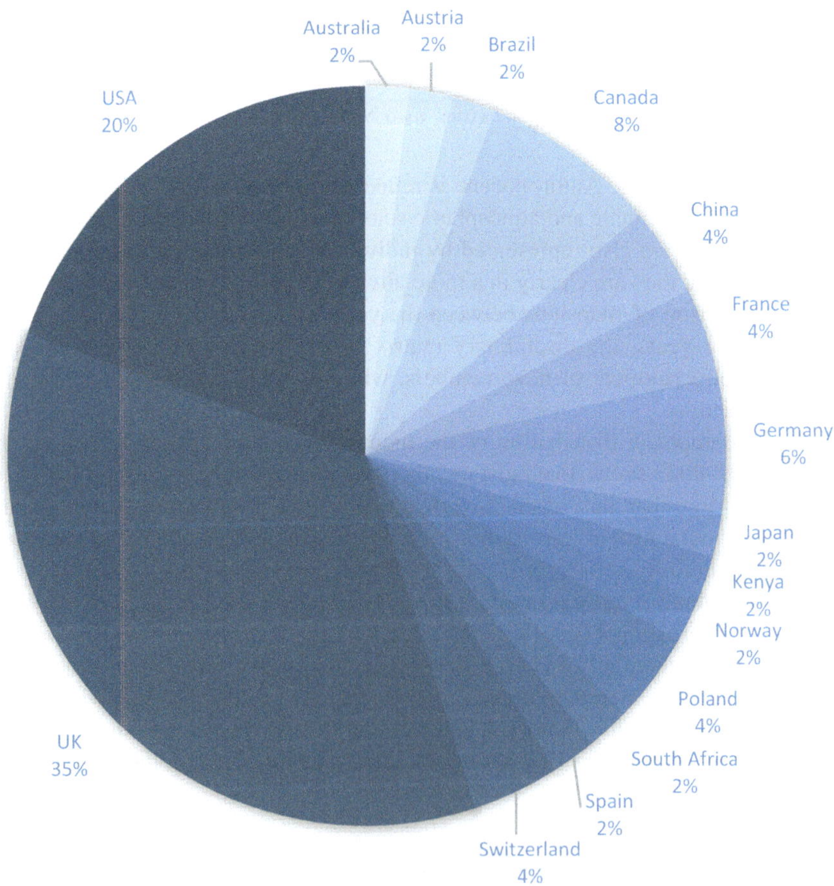

Fig. 3.3 Geographical distribution of AWG membership

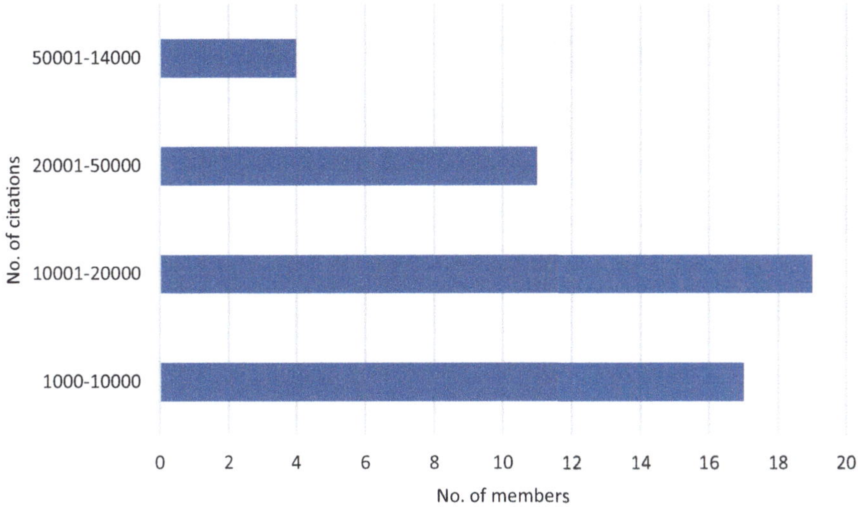

Fig. 3.4 Numbers of AWG members according to their citation numbers

thus, reinforces the homophily and homogeneity of a group. However, it should be noted that some AWG members joined out of their own initiative, based on their thematic interest in the Anthropocene (Will 2021, p. 114). Despite its skewed distribution, the AWG has managed to include a broad range of institutional and regional backgrounds.

According to the citation numbers of their publications, AWG members are active and well-cited scholars. As of December 2023, their publications have attracted between 1200 and 68,000 (Fig. 3.4).[4] One former member, late Nobel laureate Paul W. Crutzen, tops the list with almost 140,000 citations. Other indicators corroborate the impression of an active scientific community. The "h-index," which measures both the number and the impact of publications, ranges from 12 to 154 among AWG members (Fig. 3.5).

Following a rule-of-thumb interpretation, an h-index of 20 indicates a "successful scientist," 40 an "outstanding scientist," and 60 a "truly unique" scholar (Hirsch 2005). Ignoring the fact that disciplines exhibit a great deal of variation regarding their average h-index,[5] the AWG membership comprises a majority of very, even extremely well-established scientists. Due to the variation of academic age status and disciplinary differences in citation patterns, one cannot deduce an overall assessment of the AWG's collective reputation from such indicators. It is safe to say, however, that the AWG members are well integrated into their respective disciplines

[4] Citation counts and h-indices were collected from Google Scholar (where available) in November 2023.

[5] For instance, a bibliometric analysis of publications across a wide range of categories during the time period from 2010 to 2014 estimated an average h-index of 44 for palaeontology, 55 for geology, and of 155 for multidisciplinary geosciences (Patience et al. 2017).

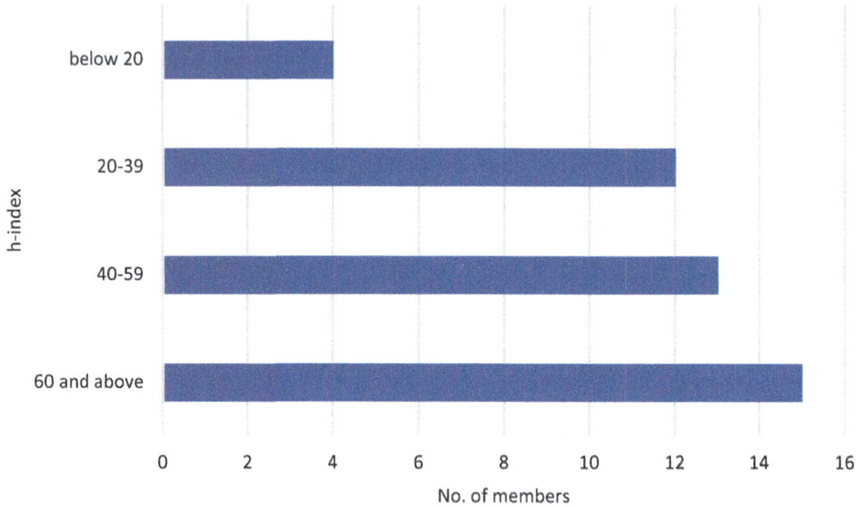

Fig. 3.5 Numbers of AWG members according to h-index categories

and are able to lend weight to the group's deliberations based on their scientific reputation.

3.3 Cooperation and Citation Patterns

To examine the AWG's scientific output and impact more closely, we collected publication data for all members for the years 2000 through 2021 from the bibliographic database *Web of Science* (WoS).[6] Formerly known as *ISI Web of Knowledge*, WoS provides an interdisciplinary collection of articles and incorporates, among others, the Science Citation Index (SCI), Social Sciences Citation Index (SSCI), and Arts and Humanities Citation Index (A&HCI). The resulting 4250 books, articles, and other publications cover the Anthropocene but also other topics; they span a period that begins before the AWG was founded and ends when the outcome of the AWG's work was still uncertain. The publications are relatively evenly distributed across the years, with a modest peak in the year 2015 and a relatively stable flow of publications since then (see Fig. 3.6). This selection allows us to observe the development of topics and citation patterns since Paul J. Crutzen and Eugene Stoermer introduced the term to a broader audience (Stoermer and Crutzen 2000; Crutzen 2002) and after the subsequent establishment of the AWG, up to and slightly beyond the AWG's vote in the year 2019 advocating a beginning of the Anthropocene in the

[6] Bibliographic data were obtained through individual searches by author names in January 2023. Limitations of the source data, regarding both the scope of indexing and the consistency of first names and initials, result in some omissions and misclassifications.

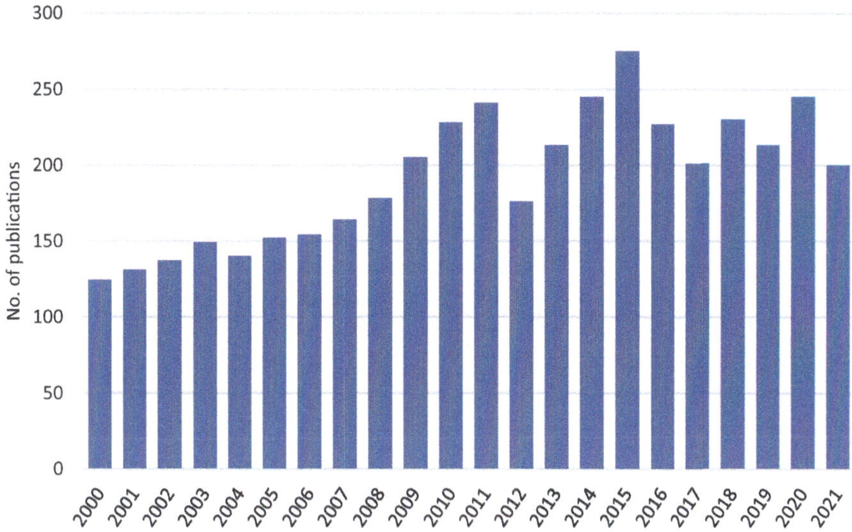

Fig. 3.6 AWG members' publications per year

mid-twentieth century. However, the publications of AWG members do not only concern the Anthropocene. Therefore, they are quite evenly distributed across the observed time period[7]:

We have already observed that AWG members are active contributors to their research fields, and the number of articles in our collection corroborates this observation. Even if publications by multiple authors are "fractionalized," i.e., discounted corresponding to the number of coauthors, most AWG members have an extensive publication record. The 10 most productive authors of the AWG are represented by 30 to almost 90 individual, fractional contributions. Considering that our sample only covers a specific time period and omits items that are not indexed by *Web of Science*, the numbers are probably near the lower bounds of actual productivity and must be interpreted with care.[8]

As aforementioned, AWG members constitute an international scientific community, with a majority of scholars working in institutions in the USA and in the United Kingdom. This is reflected in the locations of the corresponding authors, who are predominantly based in the United Kingdom and in the USA (both around 20% of publications). Researchers from China make up 12% of the corresponding authors, followed by Germany, Canada, and France. However, with a multiple-country publication ratio of well above 50%, many publications have coauthors from different countries.

[7] Unless stated otherwise, bibliometric analysis was conducted using the *bibliometrix* package in R (Aria and Cuccurullo 2017).

[8] For a more comprehensive account of the Anthropocene literature, also including other disciplines, see Brauch (2021).

Table 3.1 Most relevant journals

Journal	Number of articles
Quaternary science reviews	113
Geochimica et Cosmochimica Acta	96
Quaternary international	83
Palaeogeography Palaeoclimatology Palaeoecology	79
Science	77
Science of the Total environment	74
Earth and planetary science letters	62
Environmental Science & Technology	61
Radiocarbon	58

The perspective of the AWG is not only international but also interdisciplinary. As a group within the IUGS, however, it is primarily composed of geoscientists and researchers from adjacent disciplines. This focus is reflected in the spectrum of the most relevant journals that AWG members have published in (Table 3.1).

As a chronostratigraphic unit, the Anthropocene would be part of the Quaternary period. The AWG's immediate institutional environment, i.e., the *Subcommission on Quaternary Stratigraphy* (SQS), matches its epistemic outlook: "Quaternary Science Reviews" and "Quaternary International" claim first and second place in the ranking, respectively, while general science journals and more specialized publication outlets in geochemistry and climate change also featuring among the Top 10.

3.3.1 Scientific Collaboration

A scientific community is, first and foremost, a "knowledge community" (cf. Brauch 2021, p. 359). It addresses shared topics and questions from a disciplinary or interdisciplinary perspective. Scientific cooperation occurs in two forms: Researchers work either in teams and jointly publish results or individually, using and building on other researchers' findings. In so doing, scientists refer to other scientists' work by citing relevant publications. Therefore, patterns of coauthorship and citations are good indicators of the internal cohesion and external embeddedness of a scientific community. Multiple authorship is common in the geosciences, and the AWG is no exception. AWG members have published as coauthors with their research groups and colleagues as well as with other AWG members. A helpful way to illustrate the resulting social structure is to plot the network of coauthorships: In Fig. 3.7, authors are conceived as nodes of a graph whose edges are based on coauthorship; i.e., if two authors contribute to the same publication, they are connected. The size of the nodes represents the number of coauthorships: the more coauthorships with others, the bigger the circle.

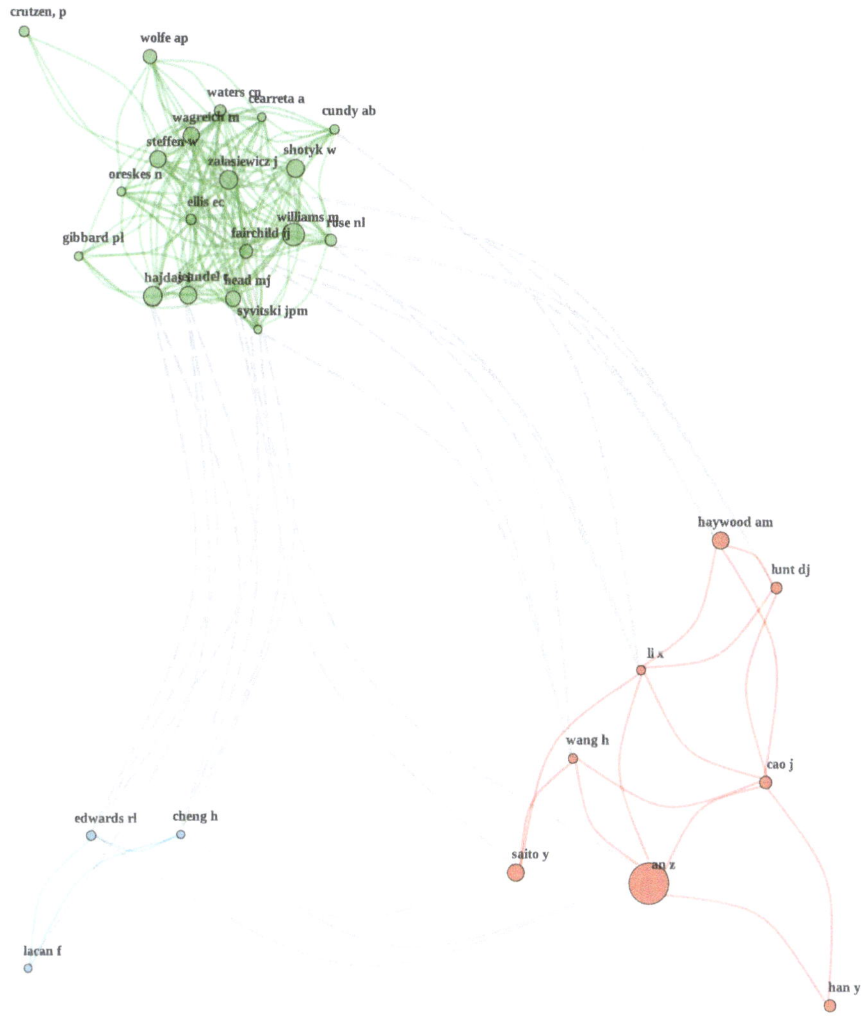

Fig. 3.7 Author collaboration (layout algorithm Fruchterman-Reingold, walktrap clustering, $n = 30$)

First of all, the pattern of coauthorships shows a high degree of collaboration. There is a large cluster of researchers who have jointly published several times, including the former and current AWG chairs Jan Zalasiewicz and Colin Waters, prominent Anthropocene protagonists from environmental and Earth systems science such as Erle Ellis and Will Steffen as well as, primarily via his collaboration with Steffen, Paul Crutzen. Another cluster includes An Zhisheng, a geologist specializing in quaternary geology, paleoclimates, and global change, and other researchers (not all of them AWG members) who focus on climate modeling and environmental change. Finally, there is a relatively small cluster of scientists

(R.L. Edwards, H. Cheng, F. Lacan) who work on climate history, ocean chemistry, and dating methods; in their publications, they collaborated with members of the other two clusters and, thereby, offer another bridge between them in addition to direct links, which exist as well. Considering the high collaboration density, particularly in the first cluster, the AWG is based not only on shared questions and methodologies but also on the social structure of a "scientific community" based on ties formed by collaboration.

This community, as we already observed with regard to the country distribution of AWG membership, is highly international yet biased toward the Anglosphere. However, the fact that more than half of the publications are written by international research teams provides another angle: If we plot the cooperation between researchers from different countries as a network of intercountry relationships, we observe a high degree of internationalization due to the many combinations in which coauthors from different countries publish together. Although the United Kingdom and the USA are the most prominent nodes of this network, Fig. 3.8 illustrates that they, as well as most other countries, are densely connected with one another.

Joint publications result from teamwork, either as a local research unit or—more often than not—as a geographically dispersed group of authors. Coauthorship based on co-presence, at least during some phase of research writing, is most commonly found in sciences that regard the "lab" as the default unit of activities, for instance, in scientific disciplines with a strong focus on experiments. It is the most obvious and, from a sociological perspective, the most consequential form of cooperation since it engages the participants in various forms of face-to-face interaction that foster the strengthening of social relationships (Goffman 1983). In contrast, scientists contributing to a joint publication might only meet occasionally or not at all. The relationships between multiple authors may, therefore, be considered "weak"

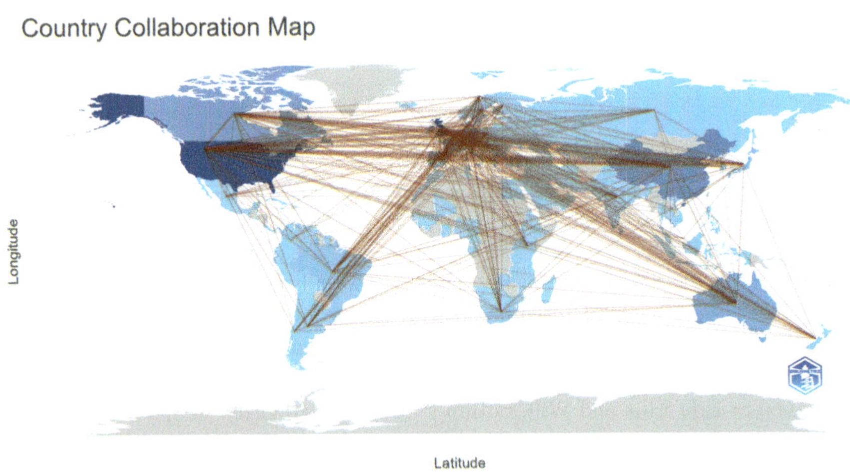

Fig. 3.8 International collaboration by countries (Only edges between countries with a minimum of three collaborations are shown; light to dark shadings depict lower to higher volumes of contributions)

rather than "strong" ties (cf. Granovetter 1973). However, joint publications without a lab background, which are the *modus operandi* of the AWG, also present research findings that must be regarded as statements of the group as a whole and are attributed to it as well as to individual researchers. They are, therefore, an essential element of establishing a scientific community.

The AWG has never been a research unit but a forum of individual researchers. Before its inaugural meeting in 2014 (see Chap. 2), its members could only meet as a side effect of another professional gathering. Based on collegial and interpersonal networks, the AWG has produced some landmark publications coauthored by multiple members. For instance, early statements such as "The Anthropocene: a new epoch of geological time?" (Zalasiewicz et al. 2011) and "A Stratigraphic Basis for the Anthropocene" (Waters et al. 2014) were coauthored, among others, by AWG members Jan Zalasiewicz, Mark Williams, Alan Haywood, and Mike Ellis. Programmatic statements leading up to the Anthropocene proposal, such as "When did the Anthropocene begin?" (Zalasiewicz et al. 2015), "The Anthropocene is functionally and stratigraphically distinct from the Holocene" (Waters et al. 2016), "Making the case for a formal Anthropocene epoch" (Zalasiewicz et al. 2017b), and "The Anthropocene as a Geological Time Unit" (Zalasiewicz et al. 2019), were clearly visible as statements of the group as even more AWG members were listed as coauthors.

3.3.2 Citations and Co-citations

A research community is, of course, based not only on cooperation but also, more indirectly, on referencing and quoting other people's works. Scientific practice is not limited to working and communicating *with* other scientists; to an increasingly large degree, it consists of communicating *about* their work in the form of citations: There is a "network of reflexive communications on top of a first-order network of social relations among scientists" (Leydesdorff 1998, p. 13). Looking at the works of AWG members that AWG members cited most frequently, the paper by Waters et al. (2016) aforementioned is among the top 10, which are listed in Table 3.2. Another familiar entry is the seminal Anthropocene paper by Crutzen (2002). The table shows both *local* and *global* citation scores (LCS and GCS, respectively): The first number indicates how many times the article is referenced in our sample (of AWG members' publications), the second number is provided by *Web of Science* and shows how many times the article is cited the entire database.

The comparison between local and global citation scores provides insights into which AWG publications are highly recognized in the (geo-)scientific community. For instance, the methodological papers by Paula Reimer et al. (with AWG member Irka Hajdas among the coauthors) provide calibration curves that are widely used tools in radiocarbon dating and have been ratified by the *International Radiocarbon Conference*. Therefore, they are frequently used and quoted in the literature. While other AWG favorites may be considered somewhat "local" heroes because their

Table 3.2 Most cited works within the sample (Local Citation Score/LCS vs. Global Citation Score/GCS)

Lead author	Year	Title	Journal	LCS	GCS
Reimer PJ	2013	IntCal13 and Marine13 radiocarbon age calibration curves 0–50,000 years cal BP	Radiocarbon	125	8643
Crutzen P	2002	Geology of mankind	Nature	90	2218
Waters CN	2016	The Anthropocene is functionally and stratigraphically distinct from the Holocene	Science	74	1021
Rockström J	2009	A safe operating space for humanity	Nature	68	7338
Wang YJ	2001	A high-resolution absolute-dated late Pleistocene monsoon record from Hulu cave, China	Science	62	2175
Syvitski JPM	2005	Impact of humans on the flux of terrestrial sediment to the global Coastal Ocean	Science	58	1776
Reimer PJ	2009	IntCal09 and Marine09 radiocarbon age calibration curves, 0–50,000 years cal BP	Radiocarbon	58	3955
Wang YJ	2005	The Holocene Asian monsoon: Links to solar changes and North Atlantic climate	Science	46	1764
Dowsett H	2010	The PRISM3D Paleoenvironmental reconstruction	Stratigraphy	46	166
Saito Y	2001	The Huanghe (Yellow River) and Changjiang (Yangtze River) deltas: A review on their characteristics, evolution and sediment discharge during the Holocene	Geomorphology	42	370

global citation rates are a bit lower, for instance, the paper aforementioned by Waters et al. (2016), the high Global Citations Score of "A Safe Operating Space for Humanity" by Rockström et al. (coauthored, among others, by Will Steffen and Paul Crutzen) indicates that this paper has also been frequently cited in the general literature. We will return to the works of AWG members that strongly impacted the broader scientific community below.

Another way to look at the cohesion of a scientific community established by its citation patterns is the analysis of "co-citation" linkages: If two works share a citation, say, to Waters et al. (2016), this signals some kind of thematic proximity (Small 1973; White and McCain 1998). To be sure, citing a publication does not necessarily mean endorsing it, so one paper may quote the source approvingly while the other does so in order to criticize it. Nonetheless, sharing a reference is evidence of a shared concern and, therefore, a valid indicator of a link between two papers. This approach has obvious limitations—regarding both the data quality of "cited references" in databases such as *Web of Science* and the interpretative paucity of mere citations—but it sheds some light on the patterns of mutual acknowledgement in a research field.

A partial co-citation network of the AWG corpus, limited to the 30 most prominent nodes, is depicted in Fig. 3.9a. Once again, there is a well-connected cluster (Fig. 3.9b) featuring some previously mentioned publications such as Crutzen (2002), Rockström et al. (2009) and Waters et al. (2016). Since the citations include works outside of our observation period and the AWG membership, there are some interesting links to additional Anthropocene literature. For instance, Ruddiman (2003) is frequently cited—not necessarily in agreement—for his argument that the anthropogenic era began much earlier than Crutzen (2002) and others would have it, namely with changes in the atmospheric concentrations of CO_2 and other gases thousands of years ago due to early agricultural practices, such as deforestation and the cultivation of rice paddies. Also prominent in this cluster are an influential article from the journal *Paleoceanography and Paleoclimatology* by Lisiecki and Raymo (2005), which provides insights into the Pliocene-Pleistocene climate history and thereby into long-term climate change, and a long-ranging review of global climate change by Zachos et al. (2001) published in the journal *Science*.

A final angle on the citation patterns underlying the Anthropocene debate traces the temporal ordering of references. The plot of the historical citation network in Fig. 3.10 illustrates not only that Crutzen (2002) was a truly seminal contribution but also that it took quite a few years to gain a foothold in the scholarly literature. Furthermore, there is an increase in both publication activity and mutual referencing after 2009. This hints at the epistemic effects of the institutionalization of the AWG: Apparently, it boosted both the production of pertinent research and the networking among research results. Probably due to the incomplete indexing of literature toward the end of the period of observation, the two last entries in the chart date from 2017 to 2018, two publications that draw on the previous contributions in different ways: On the one hand, dealing with the past, Walker et al. (2018) report on the formal ratification of a tripartite division of the Holocene; on the other hand, concerning the present and pointing toward the future, the AWG presents its interim summary of evidence (Zalasiewicz et al. 2017a).

The patterns of citation and co-citation of AWG members during the time period between 2000 and 2021 demonstrate that the AWG is not only an organizational or "bureaucratic" body (see Chap. 2) but also an epistemic or "knowledge" community (Brauch 2021, p. 359). Membership in the AWG is, of course, an indicator or pertinent interest and relevant research expertise. Against this background, the institutionally focused debate and networking during conferences and meetings enabled by the AWG seem to have also galvanized more and increasingly connected research activities of core members of the AWG. These activities have been carried out by an international and interdisciplinary community of researchers with a marked focus on Anglo-Saxon institutions and the geosciences.

a

b

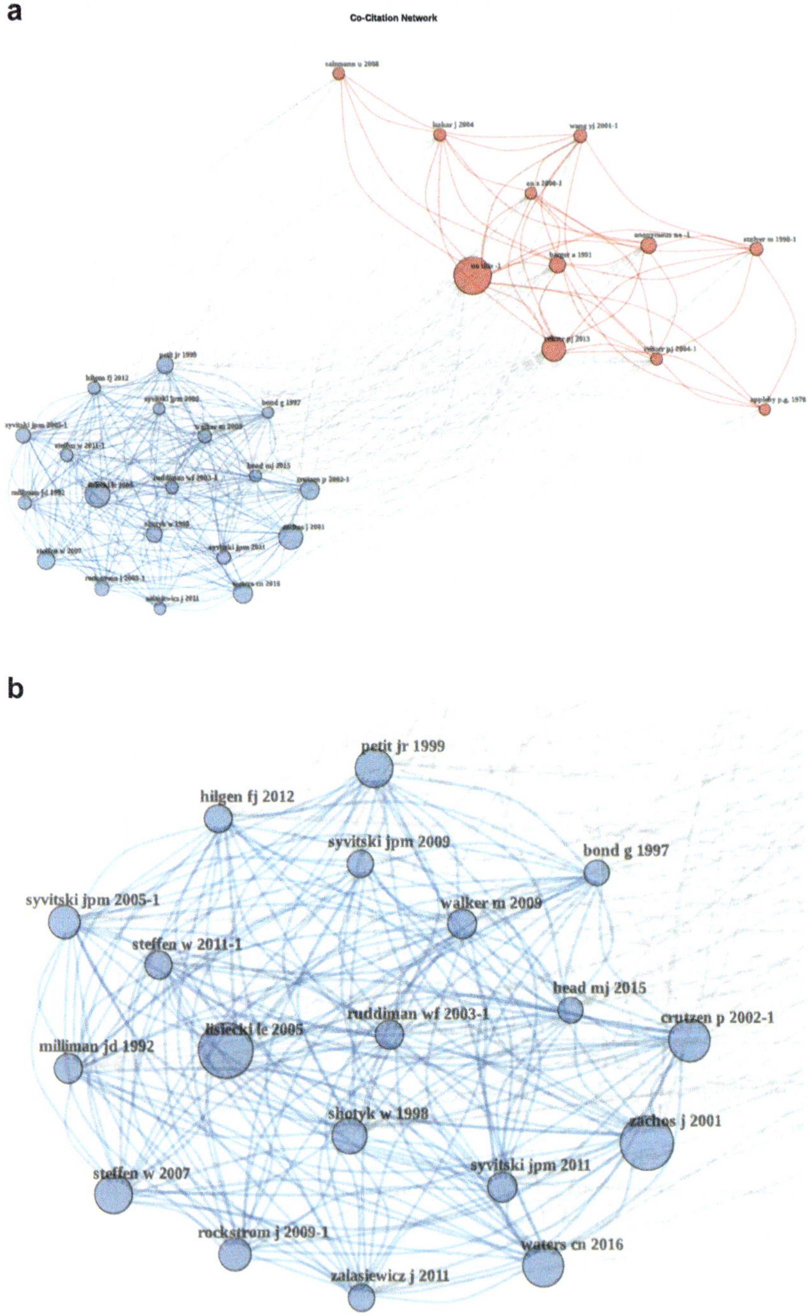

Fig. 3.9 (**a**) Co-citation network ($n = 30$); (**b**) detail of co-citation network

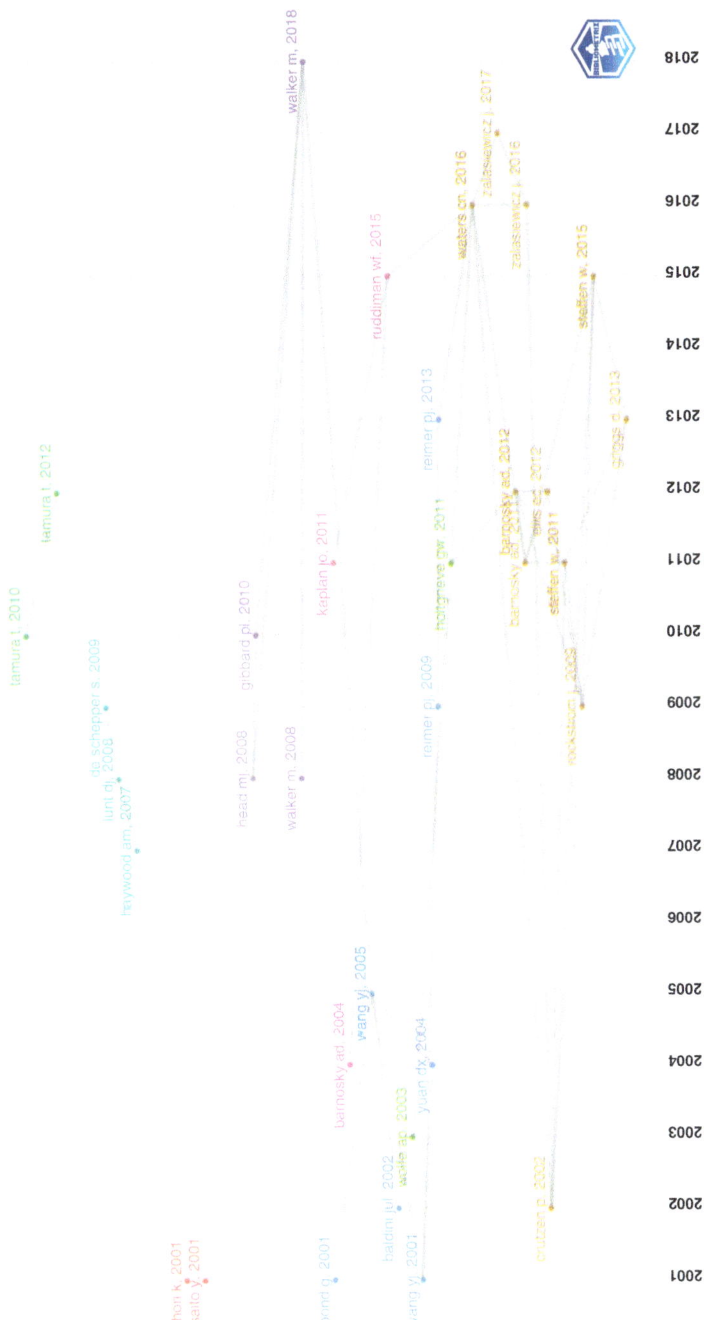

Fig. 3.10 Historical citation network

3.4 The Anthropocene as a Topic in AWG Members' Publications

The AWG set out to give the widely used notion of the Anthropocene a more precise and definite meaning. In the context of this mission, AWG members could contribute various disciplinary perspectives and expertise from their specialized research fields. However, for most of them, the Anthropocene proper is not the only, maybe not even the primary area of research. Although only 219 of the articles mention the "Anthropocene," "anthropogenic climate change," and similar terms in their titles (and even fewer explicitly use the term as one of their keywords), many present hypotheses and findings that bear on the question of how to establish the human impact on planet Earth. The list of the 10 most relevant keywords chosen by the authors (Table 3.3) gives a first glimpse of the topics addressed in the AWG members' publications.

Apart from the "Anthropocene" term, other keywords highlight questions of stratigraphy and historical as well as contemporary climate change. Since most advocators of the Anthropocene suggest terminating the Holocene epoch, the debate is also pertinent to Holocene research. Furthermore, in the public debate, the Anthropocene is commonly associated with anthropogenic climate change. Although this aspect of the concept may eventually not be decisive from a geological perspective, aspects of contemporary climate change and historical climate variations and cycles are also frequently addressed.

How has the scientific output of AWG members changed over time? We would expect that the relevance of the Anthropocene as a topic is boosted by the establishment of the AWG and the subsequent engagement of AWG members in the Anthropocene debate. Vice versa, scientists working on Anthropocene-related issues could have joined or been invited based on their contributions. Against the backdrop of research interests that either predate the Anthropocene question or are pursued simultaneously, the distribution of documents' keywords across time highlights that some topics have remained popular during the whole period of observation, while others have appeared only since 2009: The Holocene and paleoclimate,

Table 3.3 Author keywords

Keyword	# Occurrences
Holocene	105
Climate change	87
Anthropocene	82
Paleoclimate	73
Climate	57
Trace elements	54
Biostratigraphy	43
Stratigraphy	42
Diatoms	41
Geochemistry	40

for instance, have been important research topics from early on, while the Anthropocene enters the list in the year 2014, reflecting the growing relevance of the concept in the geoscientific and interdisciplinary discourse.

The most frequently used keywords in each year are shown in Fig. 3.11. The lines represent the years when that word was among the top 10 keywords. The bubble on the line highlights the year that word was used most frequently, and its size represents the frequency of use for each term. The trajectories of trending topics show that the Anthropocene became a very prominent focus of AWG members' publications after 2013, which could indicate either that members have published more extensively on the topic since then (i.e., since the establishment of the AWG) or that the composition of the AWG has changed to include more members with a focus specifically on the Anthropocene. In either case, the keywords corroborate that AWG members work in a variety of fields of geoscience, and many of them have had long-standing interests in questions of stratigraphy and/or climate change. It appears logical that such scholars should be interested in giving a definite answer to the Anthropocene questions. The Anthropocene may be the latest trend in AWG

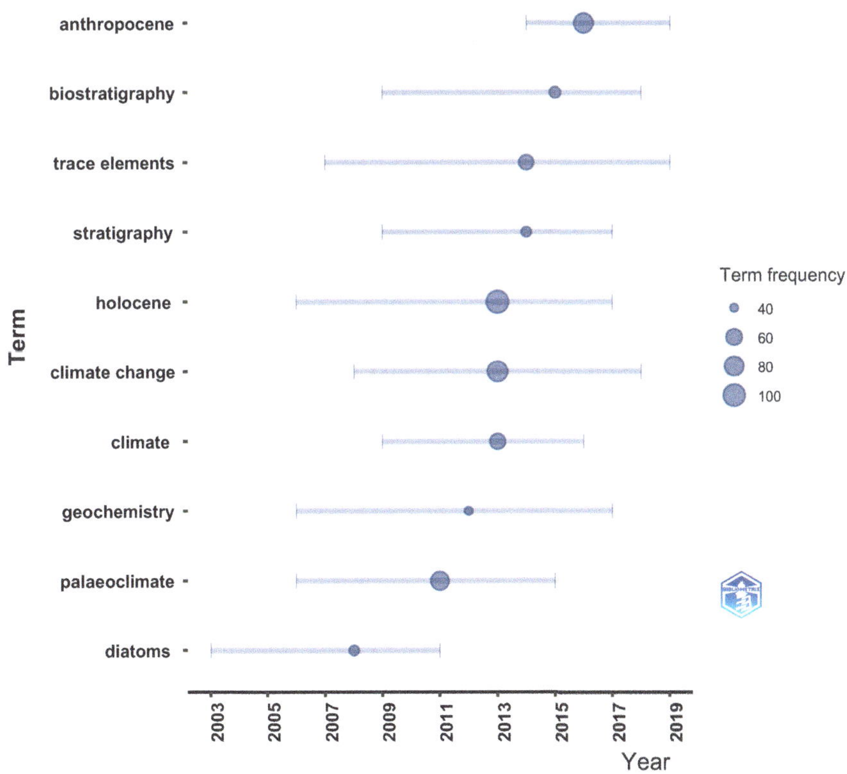

Fig. 3.11 Keywords of publications by year (minimum word frequency = 40; words per year = 10)

members' publications and driven by AWG membership, but it has been based on cognate fields of expertise.

The analysis of keyword co-occurrences reveals at least two groups of publications. In Fig. 3.12, keywords mentioned in the same publication are connected through an edge in the graph. The layout arranges keywords that are more strongly connected near to one another and, thereby, shows that "holocene," "(palaeo-)climate," and a few other keywords make up a cluster that is distinct from another one that connects the "Anthropocene" with "climate change," "palaeolimnology," "diatoms," and "trace elements."

While questions of climate and climate change indicate some overlap between the two clusters, it is interesting to note the more direct association of "climate change" with the Anthropocene as well as the significance of paleolimnology (i.e., the reconstruction of past environments of inland waters using sediment records and biological markers) and diatoms, a type of algae. This corresponds with the eventual recommendation of the AWG to choose Lake Crawford, Ontario, as the most promising candidate for the Anthropocene GSSP (McCarthy et al. 2023). However, this summary image for the whole period between 2000 and 2021 could hide thematic shifts that have occurred during this long period, which also includes the establishment of the AWG. A split into two time periods, 2000–2010 (Fig. 3.13) and 2011–2021 (Fig. 3.14), pre-AWG and AWG, respectively, indeed shows that the Anthropocene has only occupied a prominent place in the latter period.

Since 2011, and thus following the establishment of the AWG, the Anthropocene has been at the center of a cluster of keywords that is distinct from but connected to two other clusters: The first one focuses on the Holocene and questions of climate change, while the second one collects topics from stratigraphy, geochemistry, and different approaches to reconstruct Earth's past. Due to its position at the center of the graph, the climate change cluster mediates between the (more recent) Anthropocene question and the (more established) fields of studying the geo-history of the Earth.

The rise of the Anthropocene topic can also be observed in terms of the "thematic evolution" of the research field. According to Cobo et al. (2011), research themes may fall into one of four categories: highly developed, emerging or declining, basic, and "motor" themes.

The latter kind of research themes, situated in the upper-right quadrant in Fig. 3.15, exhibit robust development and are important for shaping the research domain. These "motor-themes" are both central in the network of co-occurrences and densely interwoven with other themes. In contrast, themes in the upper-left quadrant lack significant external ties to other themes despite strong internal connections; therefore, they are often highly specialized but of marginal significance within the field. Themes located in the lower-left quadrant are characterized by weak development and marginal relevance, characterized by lower density and centrality; this quadrant represents either emerging or diminishing themes. The lower-right quadrant collects themes of growing importance for the research field, which are not yet fully developed.

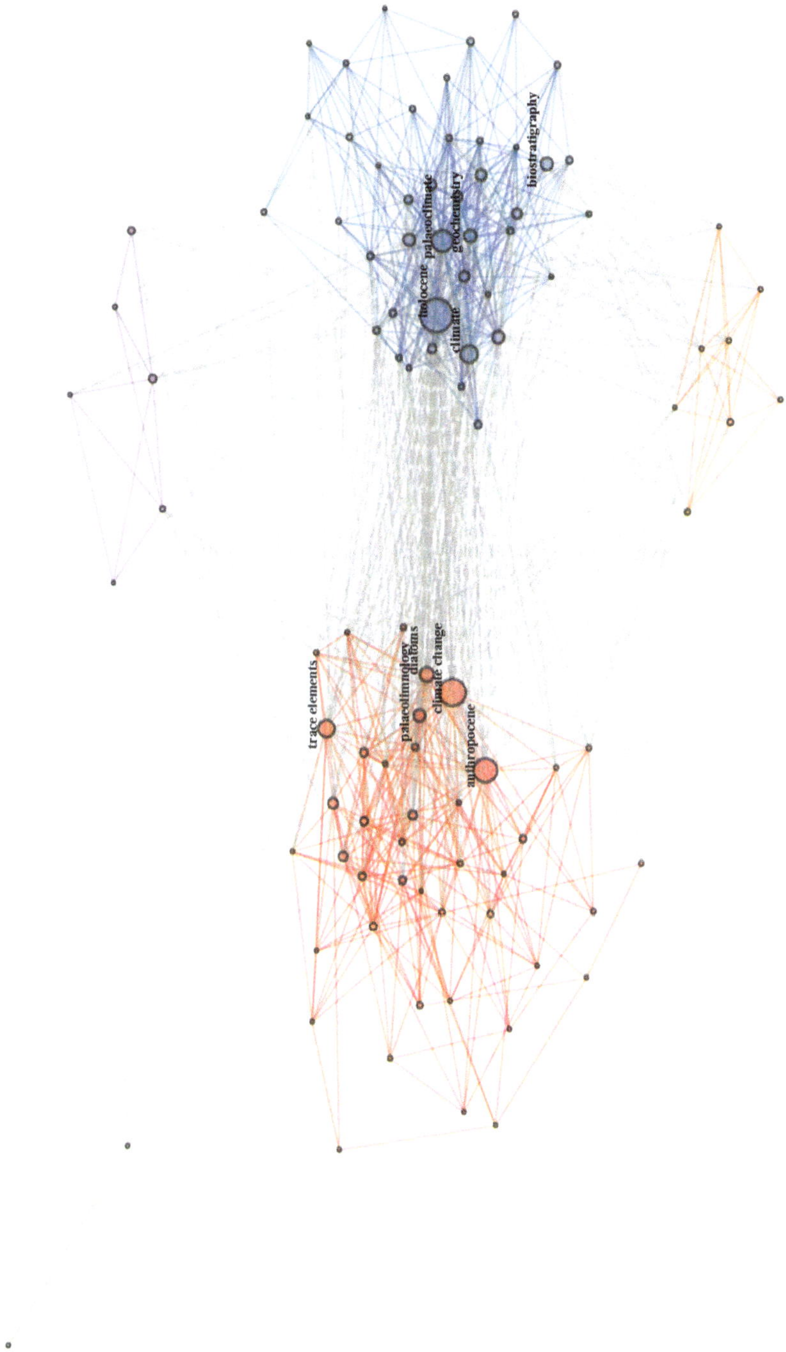

Fig. 3.12 Co-occurrences of keywords (network layout according to Fruchterman-Reingold)

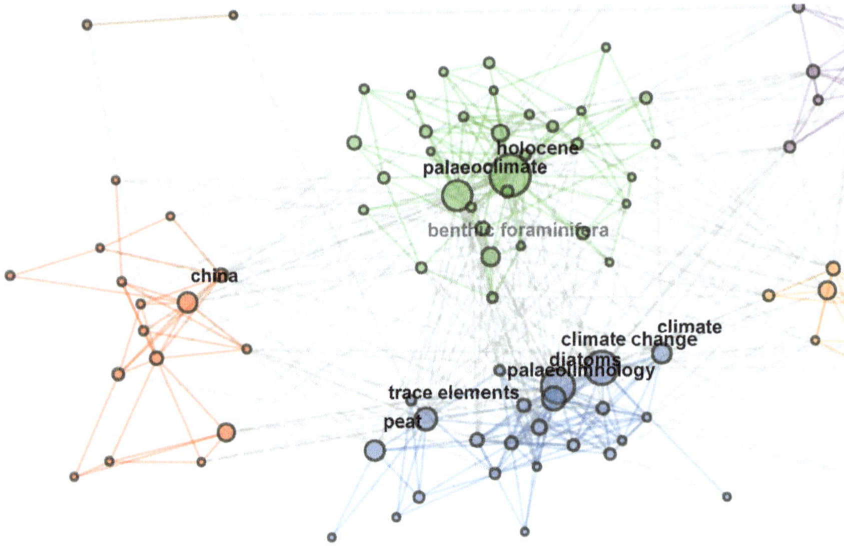

Fig. 3.13 Co-occurrences of keywords (2000–2010, detail of complete graph)

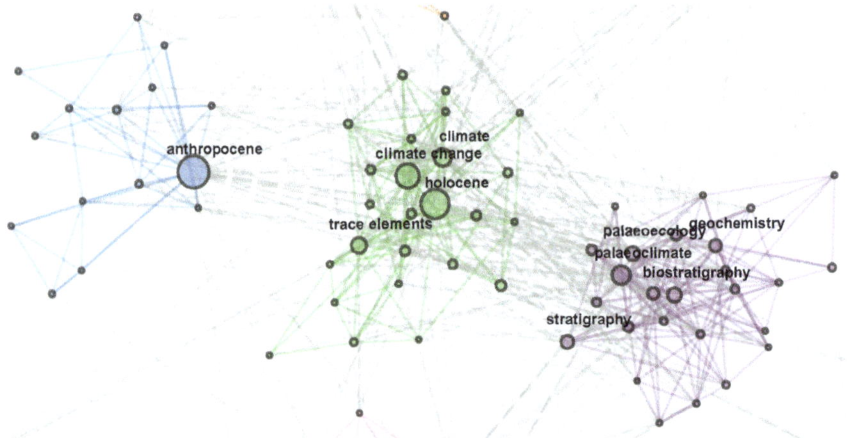

Fig. 3.14 Co-occurrences of keywords (2011–2021, detail of complete graph)

Figure 3.16 depicts the classification of research themes for the time periods pre-AWG (a: 2000–2009), after the establishment of the AWG (b: 2010–2015), and finally (c), the years after 2016, following the AWG's decision to consider the Anthropocene as terminating the Holocene in the mid-twentieth century and to define it by a "golden spike" (GSSP) using stratigraphic markers such as radioactive fallout and radiocarbon (Zalasiewicz et al. 2017a). The more detailed periodization shows that the Anthropocene had not been a major topic before the AWG was installed, but the research areas that would be relevant for the AWG's mission had

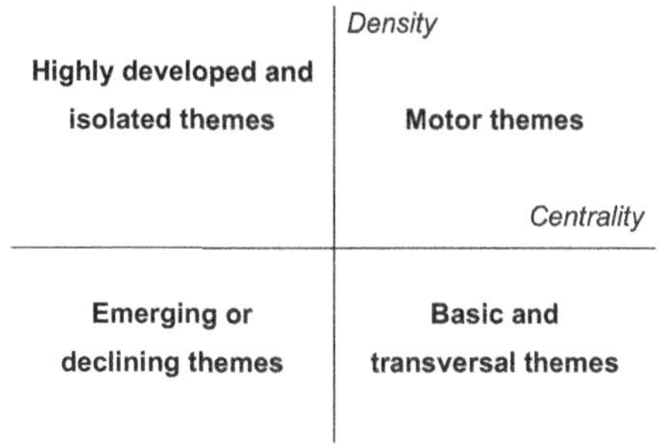

Fig. 3.15 Diagram of characteristics of research themes (Cobo et al. 2011, p. 151)

been present and central, e.g., the Holocene and climate change (12a). During the AWG's initial phase, the Anthropocene entered the picture and a more generalized interest in climate, including historical periods, emerged along with it (12b). In the final phase since 2016, the Anthropocene has become a "motor-theme" alongside radiocarbon dating, which was considered as a potential method for establishing the onset of the Anthropocene (12c).

How did research themes transform in this process? Research themes are not closed entities but combinations of interests and questions, expressed in keywords assigned by authors. Based on how combinations of keywords change and realign, we can reconstruct how research themes gradually include new topics and shed others. If research themes have some elements in common, there exists a "thematic nexus" between them (Cobo et al. 2011, p. 151). A mapping of the thematic evolution following the linkages of research themes across the time periods aforementioned is provided in Fig. 3.17.

Some groups of research themes, which consist of clusters of co-occurring keywords, span multiple time periods and thereby constitute "thematic areas." For instance, the "Holocene" constituted such a thematic area from 2000 to 2015, and "black carbon" from 2000 to 2021. Through shared keywords, thematic areas can also be linked to each other through a thematic nexus. For instance, the "Holocene" in the subperiod 2010–2015 apparently shared some issues with the "Anthropocene" of the subsequent subperiod. If a thematic area is only present in one subperiod but is linked to another thematic area in a different subperiod—for instance, "China" after the year 2010—its constituent keywords have not necessarily disappeared but have most likely become associated with a new theme.

Again, this reconstruction of the thematic evolution in the corpus shows that prior to the formation of the AWG, its members had been active in research fields such as (bio-)stratigraphy, archaeology, and most prominently, the Holocene epoch and climate change that subsequently proved instrumental to defining the Anthropocene problem from a geoscientific perspective. The latter's prominence in publications has somewhat declined since 2009, with the Anthropocene concept and related questions of deposits, trace elements, and dating gaining significance. During the last period, the Anthropocene's thematic nexus is still prominent and has linkages to more diversified and specialized fields of inquiry, such as biodiversity and depositions in various forms of soil and in the marine environment.

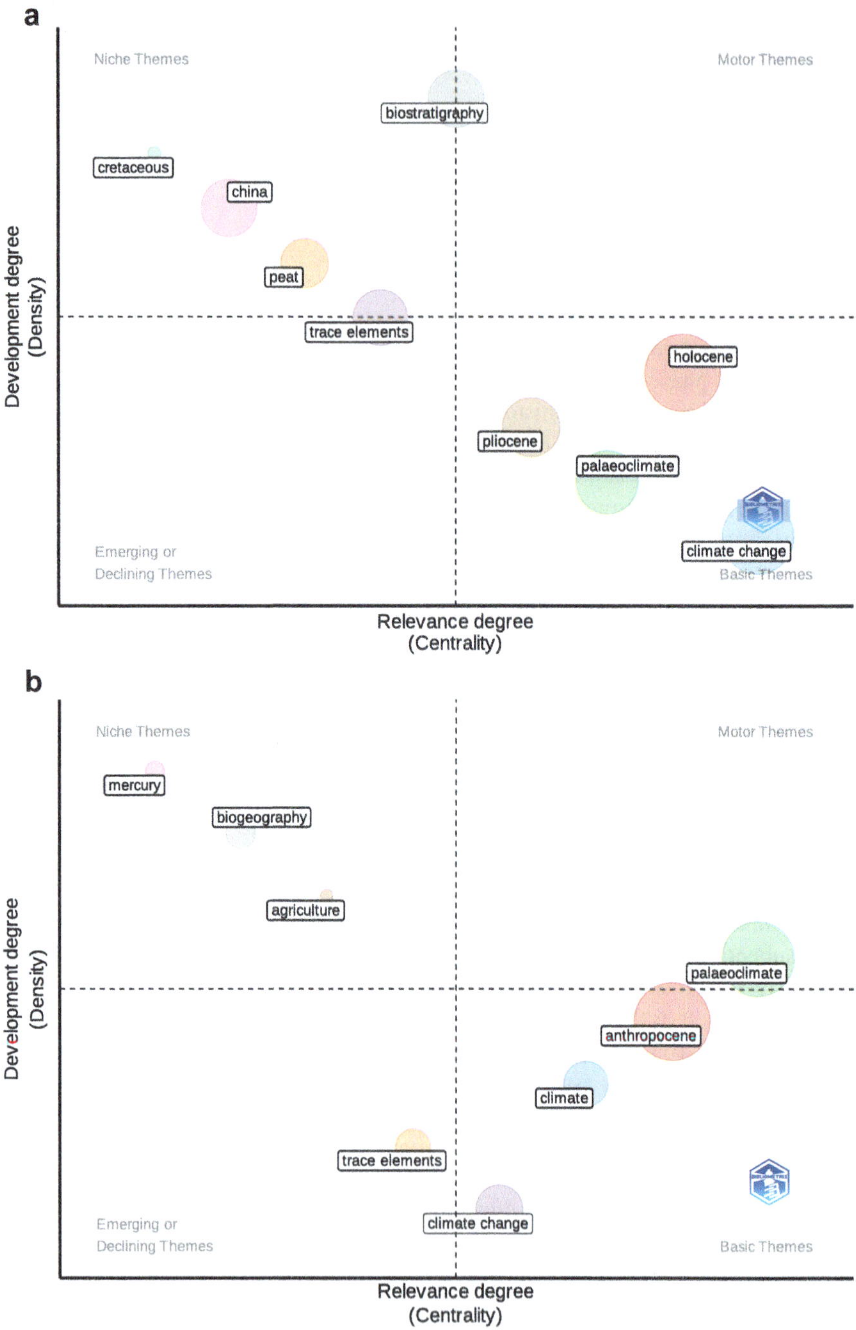

Fig. 3.16 Thematic evolution in three time periods: (**a**) 2000–2009, (**b**) 2010–2015, (**c**) 2016–2021

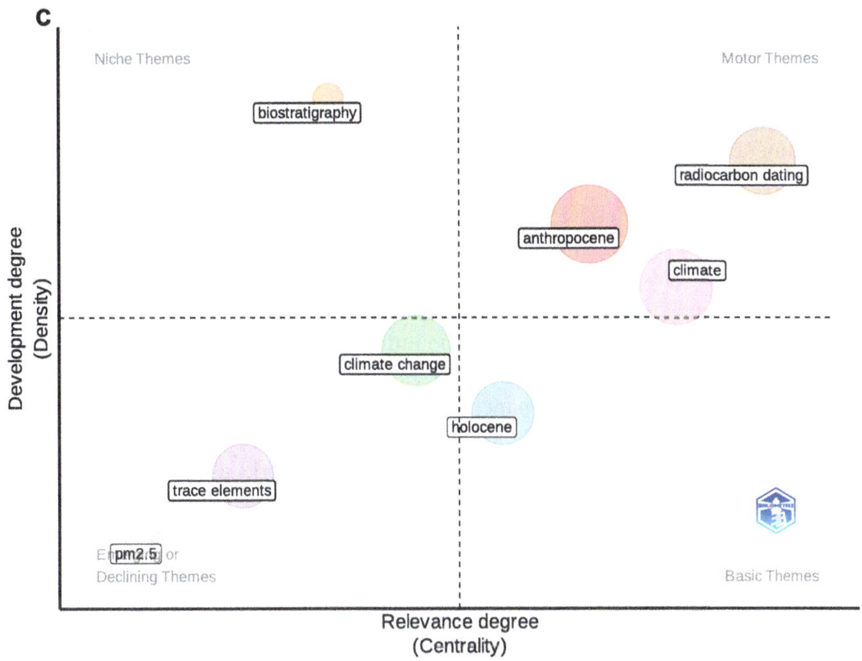

Fig. 3.16 (continued)

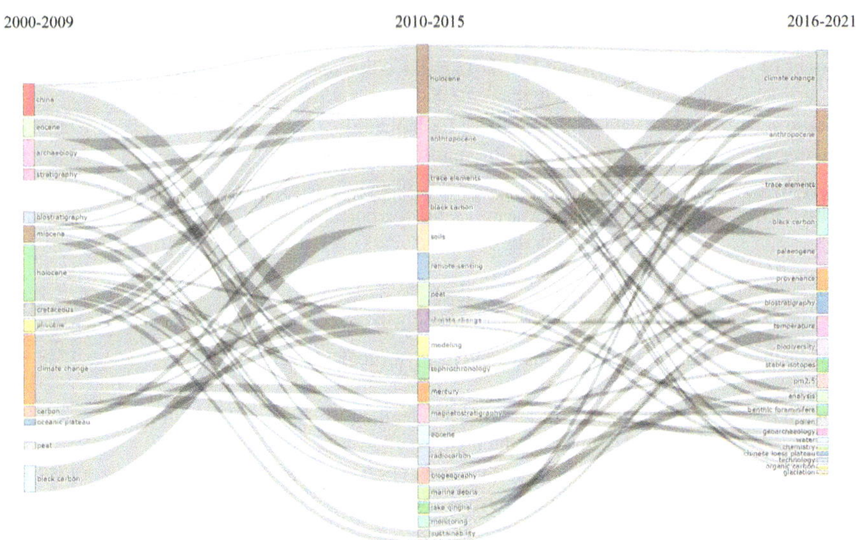

Fig. 3.17 Thematic evolution based on authors' keywords, 2000–2010–2016-2021

3.5 Impact

In AWG members' publications, the Anthropocene has become more prominent as a topic and also more densely connected with other topics. While the institutional environment of the AWG is defined by its embeddedness into the hierarchical levels of the *International Commission on Stratigraphy* (ICS) and the *International Union of Geological Sciences* (IUGS) analyzed in Chap. 2, its epistemic environment is constituted by the geological sciences and Earth Systems Science but also by the more inter- and transdisciplinary debate examined in Chap. 4. The AWG's mission has been to evaluate the potential of amending the Geological Time Scale, which ultimately required informing, and maybe influencing, the decision-making within the IUGS. However, the success of such an endeavor also depends on epistemic authority and, thus, on the resonance of the Anthropocene in the (geo-)scientific community. This section evaluates some indicators of such a resonance.

The most cited papers give an impression of which works associated with AWG members have had the greatest scientific impact (Table 3.4). The majority of them were published in *Nature* and *Science.* The citation numbers reflect citations in the whole *Web of Science* collection. Besides two methodological papers on radiocarbon dating by Reimer et al. (2009, with an update in 2013; coauthored by AWG member Irka Hajdas), the list contains the seminal Anthropocene paper on the "Geology of mankind" by Crutzen (2002) and two papers about boundaries to the

Table 3.4 Most cited publications, lead author (*) and/or coauthors AWG member(s)

Lead author	Year	Title	Journal	Citations
Reimer PJ	2013	IntCal13 and Marine13 radiocarbon age calibration curves 0–50,000 years cal BP	Radiocarbon	8643
Rockström J	2009	A safe operating space for humanity	Nature	7338
*Steffen W	2015	Planetary boundaries: Guiding human development on a changing planet	Science	4963
Reimer PJ	2009	IntCal13 and Marine13 radiocarbon age calibration curves 0–50,000 years cal BP	Radiocarbon	3955
Huang RJ	2014	High secondary aerosol contribution to particulate pollution during haze events in China	Nature	2883
Ramanathan V	2001	Aerosols, climate, and the hydrological cycle	Science	2748
Lambin EF	2001	The causes of land-use and land-cover change: Moving beyond the myths	Global Env. Change	2421
Bond G	2001	Persistent solar influence on North Atlantic climate during the Holocene	Science	2258
*Crutzen PJ	2002	Geology of mankind	Nature	2218
Wang YJ	2001	A high-resolution absolute-dated late Pleistocene monsoon record from Hulu cave, China	Science	2175

human impact on natural processes authored or coauthored by AWG members Paul Crutzen and Will Steffen (Rockström et al. 2009; Steffen et al. 2015).

The fact that only two publications are attributed to AWG members as lead authors is, to some degree, coincidental since the coauthorship order does not always reflect the factual contributions of individual authors. However, only three of the 10 publications directly address the Anthropocene question: Rockström et al. (2009), Steffen et al. (2015), and Crutzen (2002).

In order to assess which publications authored by AWG members have made an impact in the geosciences and other disciplines, another collection of articles was drawn from *Web of Science*. Covering the same time period, 2000–2021, all publications mentioning the Anthropocene in their title or keywords were selected, which resulted in a sample of 3522 references. The growing interest in the topic is reflected in the cumulative number of articles, with a take-off around 2010 and accelerating growth rates since 2015 (Fig. 3.18).

The growing interest in the Anthropocene question was, if not driven, then at least fuelled by the AWG and its members' publications since 2009. These publications served as important points of reference in the unfolding, multidisciplinary debate. Compared with the primary publication outlets of AWG members, which

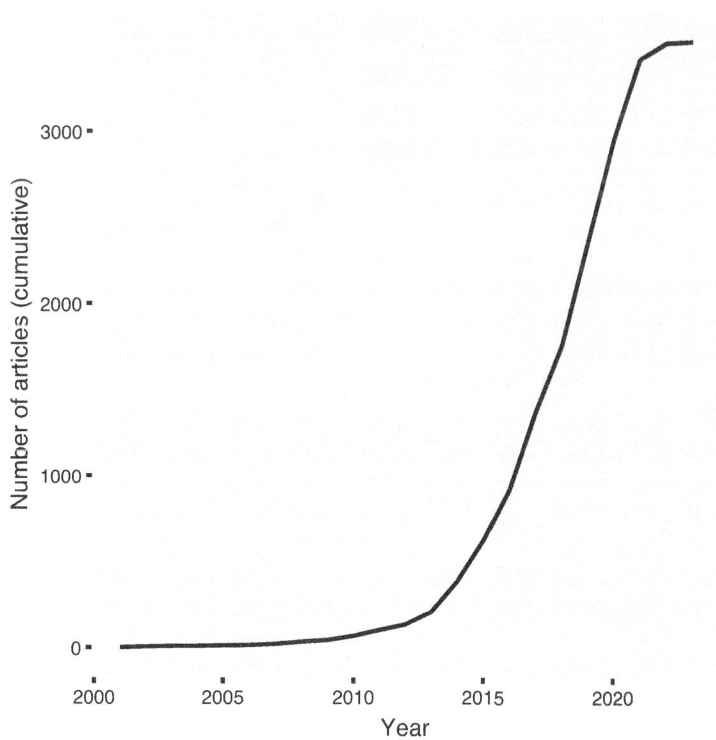

Fig. 3.18 Cumulative growth of Anthropocene-related publications in *Web of Science*, 2000–2021

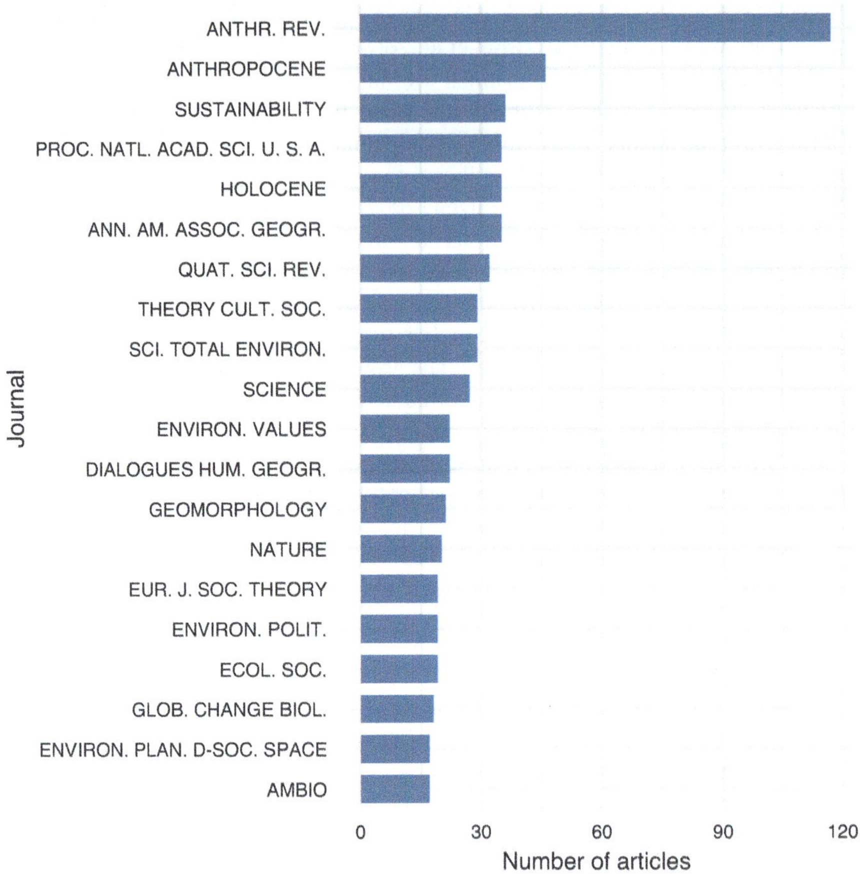

Fig. 3.19 Journal sources of Anthropocene-related publications in *Web of Science*, 2000–2021

had a strong disciplinary focus, the more general Anthropocene discourse is distributed over a more diverse set of journals (Fig. 3.19). While most journal articles are published in programmatically related interdisciplinary journals (e.g., *The Anthropocene Review*, *Sustainability*, *Anthropocene*) or in geological science and general science journals (*Geoforum, Quaternary Science Reviews*, *Science,* and *Nature*), there is also the social science journal *Theory, Culture & Society* among the top 20.

In order to shed more light on the degree of diversity of the topics discussed in the general sample of publications, automated quantitative text analysis was employed. The method of "topic modelling" allows identifying thematical foci based on the descriptions in the abstracts of the publications. This is not a hermeneutic but a "bag of words" approach, i.e., the algorithm identifies statistical co-occurrences of words that are aggregated to topics based on the statistical approach

Table 3.5 Topics of Anthropocene-related publications in *Web of Science*, 2000–2021

Highest probability terms	Interpretation
Topic 1: Earth, geological, human, epoch, system, time, global, holocene, now, great, stratigraphic, proposed	*Epoch and stratigraphy*
Topic 2: Anthropogenic, soil, records, sediments, similar, sediment, holocene, record, lake, century, human, sites	*Sediments and markers*
Topic 3: Change, systems, climate, sustainability, environmental, global, challenges, system, science, sustainable, future, framework	*Climate change and sustainability*
Topic 4: Urban, cities, climate, city, change, rise, area, natural, urbanization, papers, events, examines	*Climate change and urbanization*
Topic 5: Species, diversity, change, forest, plant, climate, population, habitat, human, populations, extinction, biodiversity	*Biodiversity and habitats*
Topic 6: Political, environmental, history, concept, politics, historical, nature, human, change, future, climate, ecological	*Environmental politics and philosophy*
Topic 7: Human, natural, humans, landscape, anthropogenic, processes, landscapes, ecological, evolutionary, effects, fire, environmental	*Anthropogenic environmental impact*
Topic 8: Environmental, global, development, climate, international, de, change, law, justice, rights, ecological, human	*Global environmental change and the law*
Topic 9: Human, nature, world, education, science, life, work, us, humans, climate, essay, knowledge	*Educational and ethical questions*
Topic 10: Ecosystem, conservation, biodiversity, ecosystems, management, ecological, services, coral, change, restoration, marine, reefs	*Marine ecosystems*
Topic 11: Climate, water, changes, river, sediment, carbon, change, global, anthropogenic, co2, atmospheric, ocean	*Sediments in lakes, rivers, and oceans*
Topic 12: Health, food, energy, human, production, pollution, global, agriculture, population, environmental, growth, environment	*Pollution and health implications*

of "latent dirichlet allocation" (LDA) (Blei et al. 2003). The idea is to "reverse engineer" the writing process, which turns ideas and topics into a text. Each article can thus be regarded as a conglomerate of several topics that it deals with. Topics consist of words that occur together frequently and may thus be characterized by their "most typical" words. For instance, certain words have a very high probability of occurring in the context of a certain topic. Table 3.5 lists these "highest probability" terms for the 12 topics and suggests an interpretation based on these terms and additional information about the word frequencies for each of the topics.[9]

[9] Topic modeling is an inductive procedure in which the analyst has to decide the number of topics. By comparing the coherence scores of various numbers of topics for a given text corpus, an informed choice is possible. In our case, coherence scores increase with the number of topics (starting from 3) and flatten out at around 12 topics (more topics always slightly increase the score but become increasingly difficult to interpret). The analysis was performed using R's "stm" package for "structural topic models" (Roberts et al. 2019).

The statistical nature of the LDA approach can only provide a rough overview, a kind of "distant reading" (Moretti 2013) of the Anthropocene debate that provides a structuralist perspective on which issues occur together frequently, but it is not a semantic analysis of the Anthropocene discourse. The identified topics demonstrate the broad range of questions, topics, and problems that are discussed with regard to the Anthropocene. They go beyond the geo-scientific framing of the Anthropocene, but both the substantial question of when the Anthropocene began and methodological questions of how to establish it as an empirical fact are well presented (for instance, in Topics 1, 2, 7, and 11). Since the social sciences and the humanities are not indexed as comprehensively as the natural sciences by *Web of Science*, especially when it comes to monographs and edited volumes, this analysis probably underestimates the multidisciplinary character of the Anthropocene debate, which cannot be judged on the basis of citations alone (see Chap. 4 for more details and analysis of the topics of this literature, particularly in the humanities).

However, due to its reference to changes and dynamics of the Earth system, this debate must always refer back to foundational and newly emerging perspectives from geology and the geosciences more broadly conceived. At least since 2009, the AWG has provided a convenient point of reference for such an observation of recent developments in the Anthropocene debate. We should, therefore, expect that publications of AWG are frequently cited in the general Anthropocene literature. Table 3.6, which lists the top 20 sources according to citation frequencies, shows that many of them are publications written by AWG members, as either lead or coauthors.

It is not surprising that older but also more foundational, sometimes also more general publications, such as Stoermer and Crutzen (2000) and Crutzen (2002), should be cited more frequently than more recent and strictly geological ones. There also is a reputation-based "Matthew effect" (Merton 1968) at work that reinforces the central position of well-known Anthropocene protagonists such as Paul Crutzen and Will Steffen over the years: Since their works are already frequently cited in the literature, new entries feel compelled to cite them again. Nonetheless, works by former and current chairs Jan Zalasiewicz and Colin Waters, together with large parts of the AWG membership, are also referenced by scholars beyond the AWG and beyond geology. The broader, sometimes incongruent perspective of the multidisciplinary debate is manifested in two aspects: First, critical and dissenting views on the Anthropocene, such as Ruddiman (2003), are obviously discussed more extensively; second, the Anthropocene question is linked to broader issues in social and cultural theory, considering the frequent citations of Haraway (2015) and Latour (1993). These two theorists have been instrumental in linking the Anthropocene debate not only to other disciplines but also to culture and the arts, which are discussed in Chaps. 4 and 5, respectively.

Table 3.6 Most cited sources in the general Anthropocene literature, 2000–2021

Lead author(s)	Year	Title	Journal	cited
Crutzen P*	2002	Geology of mankind	Nature	549
Stoermer E Crutzen P*	2000	The Anthropocene	IGBP global change newsletter	513
Steffen W*	2007	The Anthropocene: Are humans now overwhelming the great forces of nature?	Ambio	392
Rockström J*	2009	A safe operating space for humanity	Nature	377
Lewis SL, Maslin MA	2015	Defining the Anthropocene	Nature	318
Steffen W*	2011	The Anthropocene: Conceptual and historical perspectives	Phil. Trans. Royal Society A	311
Steffen W*	2015	The trajectory of the Anthropocene: The great acceleration	Anthropocene review	286
Chakrabarty D	2009	The climate of history: Four theses	Critical inquiry	267
Waters CN*	2016	The Anthropocene is functionally and stratigraphically distinct from the Holocene	Science	259
Steffen W*	2015	Planetary boundaries: Guiding human development on a changing planet	Science	237
Malm A, Hornborg A	2014	The geology of mankind? A critique of the Anthropocene narrative	Anthropocene review	209
Steffen W*	2011	The Anthropocene: From global change to planetary stewardship	Ambio	202
Haraway D	2015	Anthropocene, Capitalocene, Plantationocene, Chthulucene: Making kin	Environmental humanities	177
Ruddiman WF	2003	The anthropogenic greenhouse era began thousands of years ago	Climatic change	173
Zalasiewicz J*	2015	When did the Anthropocene begin? A mid-twentieth century boundary level is stratigraphically optimal	Quaternary international	153
Zalasiewicz J*	2010	The New World of the Anthropocene	Env. Science & technology	146
Zalasiewicz J*	2008	Are we now living in the Anthropocene?	GSA today	133
Barnosky AD*	2011	Has the Earth's sixth mass extinction already arrived?	Nature	124
Latour B	1993	We have never been modern	[book]	113
Crutzen P*, Steffen W*	2003	How Long have we been in the Anthropocene era?	Climatic change	112

(*) AWG members are either lead or coauthors

3.6 Conclusion

The analysis of AWG members' affiliations and their publications shows a scientific community realizing a high degree of international and interdisciplinary cooperation with a marked focus on Anglo-Saxon institutions and disciplines within the realm of the geological sciences. Membership in the AWG is, of course, already an indicator of interest in the Anthropocene question and of relevant research expertise.

Against this background, the institutional focus of the Group and the thematically focused exchange among AWG members seem to have galvanized more diverse and increasingly connected research activities of core members of the AWG. Owing to the standing and reputation of its members and their research, the AWG has left its footprint in the broader Anthropocene debate. Although that debate adds more perspectives and approaches, it cannot ignore the progress of the AWG's process toward a formal ratification of the Anthropocene as a chronostratigraphic unit. It remains to be seen how this wider debate is affected by the rejection of the AWG's Anthropocene proposal by the SQS. Still, the extant contributions of the AWG will remain important points of reference even if the debate should move on and develop new alternatives to the chronostratigraphic framing of the Anthropocene question.

References

Aria, Massimo, and Corrado Cuccurullo. 2017. Bibliometrix: An R-tool for comprehensive science mapping analysis. *Journal of Informetrics* 11 (4): 959–975.

Blei, David M., Andrew Y. Ng, and Michael I. Jordan. 2003. Latent dirichlet allocation. *Journal of Machine Learning Research* 3 (Jan): 993–1022.

Brauch, Hans Günter. 2021. The Anthropocene concept in the natural and social sciences, the humanities and law—A bibliometric analysis and a qualitative interpretation (2000–2020). In *Paul J. Crutzen and the Anthropocene: A new epoch in Earth's history*, ed. Susanne Benner, Gregor Lax, Paul J. Crutzen, Ulrich Pöschl, Jos Lelieveld, and Hans Günter Brauch, 289–438. Cham: Springer.

Cobo, M. J., A. G. López-Herrera, E. Herrera-Viedma, and F. Herrera. 2011. An approach for detecting, quantifying, and visualizing the evolution of a research field: A practical application to the fuzzy sets theory field. *Journal of Informetrics* 5 (1): 146–166. https://doi.org/10.1016/j.joi.2010.10.002.

Crutzen, Paul J. 2002. Geology of mankind. Nature 415 (6867): 23. https://doi.org/10.1038/415023a

Goffman, Erving. 1983. The interaction order. *American Sociological Review* 48:1–17.

Granovetter, Mark. 1973. The strength of weak ties. *American Journal of Sociology* 78 (6): 1360–1380.

Haraway, Donna. 2015. Anthropocene, Capitalocene, Plantationocene, Chthulucene: Making kin. *Environmental Humanities* 6 (1): 159–165. https://doi.org/10.1215/22011919-3615934.

Hirsch, J. E. 2005. An index to quantify an individual's scientific research output. *Proceedings of the National Academy of Sciences of the United States of America* 102 (46): 16569–16572. https://doi.org/10.1073/pnas.0507655102.

Knorr-Cetina, K. D. 1981. *The manufacture of knowledge. An essay on the constructivist and contextual nature of science*. Oxford: Pergamon Press.

Latour, Bruno. 1983. Give me a laboratory and I will raise the world. In *Science observed. Perspectives on the social study of science*, ed. Karin Knorr-Cetina and Michael Mulkay, 141–170. London: Sage.

———. 1993. *We have never been modern*. Cambridge, MA: Harvard University Press.

Latour, Bruno, and Steve Woolgar. 1979. *Laboratory life. The social construction of scientific facts*. Beverly Hills: Sage.

Leydesdorff, Loet. 1998. Theories of citation? *Scientometrics* 43 (1): 5–25. https://doi.org/10.1007/BF02458391.

Lisiecki, Lorraine E., and Maureen E. Raymo. 2005. A Pliocene-Pleistocene stack of 57 globally distributed benthic δ 18 O records. *Paleoceanography* 20 (1): https://doi.org/10.1029/2004PA001071.

Luhmann, Niklas. 1992. *Die Wissenschaft der Gesellschaft*. Frankfurt/Main: Suhrkamp.

McCarthy, Francine Mg, R. Timothy Patterson, Martin J. Head, Nicholas L. Riddick, Brian F. Cumming, Paul B. Hamilton, Michael Fj Pisaric, A. Cale Gushulak, Peter R. Leavitt, Krysten M. Lafond, Brendan Llew-Williams, Matthew Marshall, Autumn Heyde, Paul M. Pilkington, Joshua Moraal, Joseph I. Boyce, Nawaf A. Nasser, Carling Walsh, Monica Garvie, Sarah Roberts, Neil L. Rose, Andy B. Cundy, Pawel Gaca, Andy Milton, Irka Hajdas, Carley A. Crann, Arnoud Boom, Sarah A. Finkelstein, and John H. McAndrews. 2023. The varved succession of Crawford Lake, Milton, Ontario, Canada as a candidate global boundary Stratotype section and point for the Anthropocene series. *The Anthropocene Review* 10 (1): 146–176. https://doi.org/10.1177/20530196221149281.

McNeill, J. R., and Peter Engelke. 2016. *The great acceleration. An environmental history of the Anthropocene since 1945*. Cambridge, MA: Harvard University Press.

Merton, Robert K. 1968. The Matthew effect in science. *Science* 159:56–63.

Moretti, Franco. 2013. *Distant reading*. London: Verso.

Patience, Gregory S., Christian A. Patience, Bruno Blais, and Francois Bertrand. 2017. Citation analysis of scientific categories. *Heliyon* 3 (5): e00300. https://doi.org/10.1016/j.heliyon.2017.e00300.

Roberts, Margaret E., Brandon M. Stewart, and Dustin Tingley. 2019. Stm: An R package for structural topic models. *Journal of Statistical Software* 91 (2): 1–40. https://doi.org/10.18637/jss.v091.i02.

Rockström, Johan, Will Steffen, Kevin Noone, F. Asa Persson, Stuart Chapin, Eric F. Lambin, Timothy M. Lenton, Marten Scheffer, Carl Folke, Hans Joachim Schellnhuber, Björn Nykvist, Cynthia A. de Wit, Terry Hughes, Sander van der Leeuw, Henning Rodhe, Sverker Sörlin, Peter K. Snyder, Robert Costanza, Uno Svedin, Malin Falkenmark, Louise Karlberg, Robert W. Corell, Victoria J. Fabry, James Hansen, Brian Walker, Diana Liverman, Katherine Richardson, Paul Crutzen, and Jonathan A. Foley. 2009. A safe operating space for humanity. *Nature* 461 (7263): 472–475. https://doi.org/10.1038/461472a.

Ruddiman, William F. 2003. The anthropogenic greenhouse era began thousands of years ago. *Climatic Change* 61 (3): 261–293. https://doi.org/10.1023/B:CLIM.0000004577.17928.fa.

Small, Henry. 1973. Co-citation in the scientific literature: A new measure of the relationship between two documents. *Journal of the American Society for Information Science* 24 (4): 265–269. https://doi.org/10.1002/asi.4630240406.

SQS. 2024. Working group on the 'Anthropocene' | subcommission on quaternary stratigraphy. https://quaternary.stratigraphy.org/working-groups/anthropocene/. Last access: 5 August 2024.

Steffen, Will, Paul J. Crutzen, and John R. McNeill. 2007. The Anthropocene: Are humans now overwhelming the great forces of nature? *Ambio* 36 (8): 614–621. https://doi.org/10.1579/0044-7447(2007)36[614:TAAHNO]2.0.CO;2.

Steffen, Will, Katherine Richardson, Johan Rockström, Sarah E. Cornell, Ingo Fetzer, Elena M. Bennett, Reinette Biggs, Stephen R. Carpenter, Wim de Vries, Cynthia A. de Wit, Carl Folke, Dieter Gerten, Jens Heinke, Georgina M. Mace, Linn M. Persson, Veerabhadran Ramanathan, Belinda Reyers, and Sverker Sörlin. 2015. Planetary boundaries: Guiding human development on a changing planet. *Science* 347 (6223): 1259855. https://doi.org/10.1126/science.1259855.

Stichweh, Rudolf. 2014. Die Autopoiesis der Wissenschaft. In *Wissenschaft, Universität, Professionen: Soziologische Analysen*, 47–72. Bielefeld: transcript.

Stoermer, Eugene F., and Paul J. Crutzen. 2000. The Anthropocene. *IGBP Global Change Newsletter* 41:17–18.

van Vugt, Ingeborg. 2022. Networking in the republic of letters: Magliabechi and the Dutch Republic. *The Journal of Interdisciplinary History* 53 (1): 117–141. https://doi.org/10.1162/jinh_a_01800.

Walker, Mike, Martin J. Head, Max Berkelhammer, Svante Björck, Hai Cheng, Les Cwynar, David Fisher, Vasilios Gkinis, Antony Long, John Lowe, Rewi Newnham, Sune Olander Rasmussen, and Harvey Weiss. 2018. Formal ratification of the subdivision of the Holocene series/epoch (quaternary system/period): Two new global boundary Stratotype sections and points (GSSPs) and three new stages/subseries. *Episodes* 41 (4): 213–223. https://doi.org/10.18814/epiiugs/2018/018016.

Waters, Colin N., Jan Zalasiewicz, Colin Summerhayes, Anthony D. Barnosky, Clément Poirier, Agnieszka Gałuszka, Alejandro Cearreta, Matt Edgeworth, Erle C. Ellis, Michael Ellis, Catherine Jeandel, J. R. Reinhold Leinfelder, Daniel McNeill, B. de Richter, Will Steffen, James Syvitski, Davor Vidas, Michael Wagreich, Mark Williams, An Zhisheng, Jacques Grinevald, Eric Odada, Naomi Oreskes, and Alexander P. Wolfe. 2016. The Anthropocene is functionally and stratigraphically distinct from the Holocene. *Science* 351 (6269): aad2622. https://doi.org/10.1126/science.aad2622.

Waters, Colin N., Jan A. Zalasiewicz, Mark Williams, Michael A. Ellis, and Andrea M. Snelling. 2014. A stratigraphical basis for the Anthropocene? *Geological Society, London, Special Publications* 395 (1): 1–21. https://doi.org/10.1144/SP395.18.

White, Howard D., and Katherine W. McCain. 1998. Visualizing a discipline: An author co-citation analysis of information science, 1972–1995. *Journal of the American Society for Information Science* 49 (4): 327–355. https://doi.org/10.1002/(SICI)1097-4571(19980401)49:4<327:AID-ASI4>3.0.CO;2-4.

Will, Fabienne. 2021. *Evidenz für das Anthropozän. Wissensbildung und Aushandlungsprozesse an der Schnittstelle von Natur-, Geistes- und Sozialwissenschaften. Umwelt und Gesellschaft, Band 24*. Göttingen: Vandenhoeck & Ruprecht.

Zachos, J., M. Pagani, L. Sloan, E. Thomas, and K. Billups. 2001. Trends, rhythms, and aberrations in global climate 65 ma to present. *Science* 292 (5517): 686–693. https://doi.org/10.1126/science.1059412.

Zalasiewicz, Jan, Colin N. Waters, Colin P. Summerhayes, Alexander P. Wolfe, Anthony D. Barnosky, Alejandro Cearreta, Paul Crutzen, Erle Ellis, Ian J. Fairchild, Agnieszka Gałuszka, Peter Haff, Irka Hajdas, Martin J. Head, A. Juliana, Ivar do Sul, Catherine Jeandel, Reinhold Leinfelder, John R. McNeill, Cath Neal, Eric Odada, Naomi Oreskes, Will Steffen, James Syvitski, Davor Vidas, Michael Wagreich, and Mark Williams. 2017a. The working group on the Anthropocene: Summary of evidence and interim recommendations. *Anthropocene* 19:55–60. https://doi.org/10.1016/j.ancene.2017.09.001.

Zalasiewicz, Jan, Colin N. Waters, Mark Williams, Anthony D. Barnosky, Alejandro Cearreta, Paul Crutzen, Erle Ellis, Michael A. Ellis, Ian J. Fairchild, Jacques Grinevald, Peter K. Haff, Irka Hajdas, Reinhold Leinfelder, John McNeill, Eric O. Odada, Clément Poirier, Daniel Richter, Will Steffen, Colin Summerhayes, James P. M. Syvitski, Davor Vidas, Michael Wagreich, Scott L. Wing, Alexander P. Wolfe, Zhisheng An, and Naomi Oreskes. 2015. When did the Anthropocene begin? A mid-twentieth century boundary level is stratigraphically optimal. *Quaternary International* 383:196–203. https://doi.org/10.1016/j.quaint.2014.11.045.

Zalasiewicz, Jan, Colin N. Waters, Mark Williams, and Colin P. Summerhayes, eds. 2019. *The Anthropocene as a geological time unit. A guide to the scientific evidence and current debate*. Cambridge: Cambridge University Press.

Zalasiewicz, Jan, Colin N. Waters, Alexander P. Wolfe, Anthony D. Barnosky, Alejandro Cearreta, Matt Edgeworth, Erle C. Ellis, Ian J. Fairchild, Felix M. Gradstein, Jacques Grinevald, Peter Haff, Martin J. Head, A. Juliana, Ivar do Sul, Catherine Jeandel, Reinhold Leinfelder, John R. McNeill, Naomi Oreskes, Clément Poirier, Andrew Revkin, Daniel B. de Richter, Will Steffen, Colin Summerhayes, James P. M. Syvitski, Davor Vidas, Michael Wagreich, Scott Wing, and Mark Williams. 2017b. Making the case for a formal Anthropocene epoch: an analysis of ongoing critiques. *Newsletters on Stratigraphy* 50 (2): 205–226. https://doi.org/10.1127/nos/2017/0385.

Zalasiewicz, Jan, Mark Williams, Alan Haywood, and Michael Ellis. 2011. The Anthropocene: A new epoch of geological time? *Philosophical transactions. Series A, Mathematical, physical, and engineering sciences* 369 (1938): 835–841. https://doi.org/10.1098/rsta.2010.0339.

Chapter 4
The Anthropocene as a Trans-Disciplinary Issue

4.1 Introduction

When in 2000 atmospheric chemist Paul J. Crutzen and limnologist Eugene F. Stoermer introduced the term Anthropocene, they sparked off a scientific debate, which quickly gained momentum. Today, the Anthropocene has developed into a concept, which is discussed by numerous academic as well as public communities.

As a result, the Anthropocene concept forces actors from different disciplines to leave their comfort zones and leave behind longs established cultures of knowledge production in order to enter an active negotiation process about different anthropocenic aspects beyond disciplinary boundaries. Things that seemed unambiguous for a long time have become ambiguous—boundaries between established categories are becoming just as blurred as the epistemic boundaries between the disciplines involved. This results in an interaction between collaboration and competition, which promises to spur new approaches that have the potential to overcome existing barriers that come into play in thinking with the Anthropocene.

Using selected lines of conflict, this chapter examines the structure and the consequences of negotiation processes at disciplinary interfaces resulting from the provocative potential of the Anthropocene concept.[1] Therefore, the provocation resulting for humanities and social sciences from the debate about the Anthropocene as a geological concept[2] is to be understood as an opportunity for collaboration and

---

This chapter is authored by Fabienne Will.

---

[1] I want to thank my coauthors of this volume, Martin Bohle, Boris Holzer, and Leslie Sklair, for their valuable input and suggestions.

[2] The division into a debate about the Anthropocene as a geological concept, a cultural concept, and a societal phenomenon goes back to Helmuth Trischler and has proven to be a fruitful analytical instrument to structure the debate.

M. Bohle et al., *The Anthropocene Working Group and the Global Debate Around a New Geological Epoch*, Emerging Globalities and Civilizational Perspectives, https://doi.org/10.1007/978-3-031-85175-9_4

questioning established narratives in dealing with the Anthropocene (Trischler 2016; Will 2021). On this basis, it highlights the potential of the Anthropocene to transcend established disciplinary boundaries and, as a possible academic game changer, to lead into a new form of scientific collaboration. The Anthropocene challenges especially historians, as well as geologists as experts on temporality, to question established methods and narratives in the light of the converging temporal and spatial scales as well as the shifting spectrum of actors in the Anthropocene. The analysis will place particular emphasis on ways of dealing with those challenges at disciplinary interfaces.

As a first step, this chapter will examine the provocative potential of the Anthropocene concept in geology as well as in the humanities and social sciences, accompanied by insights on the self-perception and external perception of the AWG within the overall debate. Following this, an in-depth analysis of how representatives of the different disciplines involved in the debate deal with the challenges of bringing together firstly different temporal and spatial scales and secondly different agents, exploring the potential of interdisciplinary cooperation in the Anthropocene. In doing so, this chapter aims to highlight the downside of conflicting Anthropocene discourses and argues for an even broader interdisciplinarity of Anthropocene research practice.

The findings reported in this chapter are based on the analysis of 25 expert interviews conducted with central actors of the Anthropocene debate, both with members of the AWG and actors who are not part of the AWG, and on a comprehensive literature review for the years from 2000 to 2022 (for more details, see Will 2021, pp. 36ff.).

4.2 The Provocation of the Anthropocene

4.2.1 The Anthropocene as a Provocation to Geology

The Anthropocene is a provocation to geology that becomes evident in discussions between the Anthropocene Working Group and the wider geological community as well as within the Anthropocene Working Group itself. I use the term provocation here because the reactions and counter-reactions that determine the course of the debate on various thematic aspects of the Anthropocene concept are highly emotionally charged. Frequently, the scientists involved adopt a downright defensive stance and, in some cases, an unequivocal attitude of resistance—for example, when it comes to methodological deviations within geology. Thereby, a struggle for interpretative sovereignty is emerging, which the term provocation captures better than other terms. Even if the geological provocation results from a variety of factors, the tense debate points to being rooted in the established methodology of geological knowledge production, which is questioned in the debates about the Anthropocene.

Since the 1970s, the process of dividing geologic time follows clearly defined criteria (Gradstein 2020; Holmes 1947; Murphy and Salvador 1999; see also Chap. 2). As the number of time units defined in the Geologic Time Scale (GTS) shows, the criteria have proven to be effective. Although there have been repeated discussions about particular aspects of the stratigraphic approach, the methodology that characterizes the stratigraphic discipline as such has not been questioned so far, neither from a geoscientific point of view nor from outside the discipline (Will 2021, pp. 73ff.).

In the course of the Anthropocene debate, this is now changing. This explains why the Anthropocene can be seen as a provocation to geology on several levels. Even though it is not intended to provoke, it is indeed perceived as such in parts of the wider geological community.

While the field of stratigraphy has so far focused exclusively on enclosed, past time units, the Anthropocene, with its currently agreed beginning in the 1950s, is of high historical relevance, but, geologically speaking, almost not existent (Waters et al. 2018). The only exception for a ratified geochronological time unit that has not yet ended is the Holocene epoch with the Meghalayan as its currently last stage. However, with its beginning about 11,700 years ago, it offers a rich spectrum of stratigraphic data despite being open-ended. In terms of its thickness, the *Anthropocene Series* instead differs fundamentally from the type and nature of the series underlying previously ratified geological epochs (Walker et al. 2015; Finney and Edwards 2016; Gibbard and Lewin 2016). This becomes particularly evident when searching for GSSPs that document the onset of the Anthropocene. While some cores are quite rich in stratigraphic data and document a clear change for the time around 1950, candidate GSSPs from the deep ocean floor are less suitable, because "large parts of the deep ocean are too slow to have developed a meaningful thickness of Anthropocene sediments" (Waters et al. 2023c, p. 11). The temporal novelty of the Anthropocene as a possible geological epoch thus stems not only from its brevity but also from its future dimension (Finney and Edwards 2016; Gibbard and Lewin 2016; Smil 2015; Wolff 2014; Visconti 2014). Geologists involved in the debate suddenly face the challenge of incorporating probabilistic analyses into their considerations. Although the intended formalization of the Anthropocene was and will be based on traceable evidence in already deposited rock strata as the detailed analysis of possible GSSPs to define the lower boundary of the Anthropocene epoch and the recent decision of the AWG for the Crawford Lake candidate site as the GSSP for the Anthropocene show (Waters et al. 2023b , c) (Klingan et al. 2023), the working group cannot avoid to include possible future scenarios and lines of development of planet Earth in its considerations (Steffen et al. 2018; Steffen and Morgan 2021):

> "[I]n the formal part of the work we have to say the future stops here, and all the evidence we consider stops here" (Interview Zalasiewicz, 2018). "That's true but we have got into the realms of futurism [...] because it's impossible for us to think about defining a boundary without thinking about what might come next. And we have lived in different trajectories. We have thought about where the planet may go in the not too distant future" (Interview Williams, 2017). "We've also talked about whether signals will be stable [...] But that's not

saying that we are using the future to decide where the boundary is, it's helping to inform" (Interview Waters, 2017).

The Anthropocene—whether formalized or not—has just begun, and its location at the epoch level implies for some the assumptions that it will last for millions of years, and that the Earth systemic changes gathered under the term Anthropocene will catapult the Earth into a "no-analogue state." The indicators of this change have already been so extensive that there is sufficient reason to believe that the Earth systemic conditions thereby triggered will persist over a long time span. Whether the geological considerations on the Anthropocene are still compatible with the suggested duration for epochs of 13–35 million years as defined in the International Stratigraphic Guide (ISG) seems questionable (Murphy and Salvador 1999; Hedberg 1976). On the contrary, although the proposal to formalize the Anthropocene as an epoch has been rejected by the SQS recently (IUGS 2024; Witze 2024a, b), the concept of geological epochs seems to be undergoing a fundamental change and stimulates intense debates within geology—especially with regard to their duration, which was already hinted at with the negotiation of the Holocene (see e.g. Gradstein et al. 2005; Walker et al. 2009; Steffen and Morgan 2021).

Mainly two aspects seem to be central with regard to the change in criteria likely to occur: First, the development of new and more accurate technological dating techniques is the fundamental prerequisite for the increasingly precise dating of strata. Second, analytical results of the geologically recent past are particularly insightful because human life forms evolved in a comparatively rapid manner in geological terms. It is true that human impact on the different Earth systems did not only begin with the Anthropocene—the crucial difference in this respect is the scale of anthropogenic influence. Therefore, strata that cover a clearly smaller time unit are still rich in deposits. Both aspects indicate that the shift in the conception of geological epochs is closely related to human history. Reports from the ICS and the SQS suggest that there seems to be an awareness of this impending change within geology. The reports repeatedly recommend the establishment of a fifth sub-working group under the SQS whose task would be to examine "the utility of formal definition of short-time divisions" (Finney 2009; Gibbard 2010). Nothing less than the concept of geological deep time—since the emergence of geology a defining disciplinary characteristic—is hereby scrutinized. In the meantime, however, the suggestion to establish a new working group has disappeared from the agenda again.

Moreover, intra-geological discussions on the question if the Anthropocene meets the criteria to count as an epoch persisted despite the consensus already formulated by the AWG in 2016 as a result of a vote in the run-up to the International Geological Congress in Cape Town. Counter-proposals since then aimed at formalizing the Anthropocene as a geological event rather than an epoch (Gibbard et al. 2022; Edwards et al. 2022; Head et al. 2021, 2022). With the AWG's decision in favor of the Crawford Lake GSSP in July 2023, the submission of the formalization proposal to the SQS on October 31st, 2023 and the rejection of this proposal by the SQS in March 2024, this discussion reached a new peak (Waters et al. 2023a; Ellis 2024; Zhong 2023). It points to one of the most important questions for geology when it comes to defining time units: whether what is traceable in the strata points to an isochronous or diachronous change. While this aspect is secondary or even irrelevant for other disciplines, it forms the basis of geological time division. With the proposal to formalize the Anthropocene as an event, Gibbard et al. argue for its diachronous features and see here the following advantage: "A shift to an event framework for defining the Anthropocene [...] frees the concept from the constraints of geological formalization as well as from its alignment with established chronostratigraphical and geochronological units within the Holocene Series/Epoch [and] [...] provides a universal term (a common language) that facilitates communication beyond the geoscience community with the social sciences and humanities" (Gibbard et al. 2022, p. 398).

Interestingly, the very actors who for years have been critical about the involvement of non-geoscientific disciplines into geological debates are now arguing in favor of maintaining the interdisciplinary connectivity of the concept—which, according to Gibbard et al., would be best ensured if the Anthropocene was recognized as an event. In a recent paper, the AWG commented on the proposal to formalize the Anthropocene as an event and argued why, from their point of view, it should be formalized as an epoch:

> "We thus propose recognizing a long, slow-unfolding Anthropocene Modification Episode (the AME), outside of formal chronostratigraphy, leading to and incorporating the Great Acceleration Event Array (the GAEA) that signals the onset of a chronostratigraphic Anthropocene epoch/series" (Waters et al. 2022, p. 23). "Given the intensity, magnitude, planetary significance and global isochroneity of the GAEA, it provides a suitable level for recognition of the base of the Anthropocene as a series/epoch" (Waters et al. 2022, p. 1).

A statement published shortly after the rejection of the proposal to formalize the Anthropocene as an epoch through the SQS affirms the AWG's position:

> "Humans have had a long and complex impact on the planet, true. For almost all that time, they left their marks on Earth – but did not utterly overwhelm it. Less than a century ago, processes that began during the Industrial Revolution swung into overdrive. That's the Anthropocene as an epoch. It's real, it's already made geology, and it won't go away. Best to acknowledge it, to help us cope with its consequences" (Turner et al. 2024).

The term itself represents a further novelty to geology and as such another source of irritation. Firstly, it originates from an Earth System Science thesis, which has only been examined for its geological meaningfulness from 2000 to 2009. As a follow-up to this initial pre-institutional investigation, the AWG was established in 2009. However, if one considers the relationship between Earth System Sciences and geology, which Will Steffen describes as "symbiotic and well defined," the ESS origin of the concept is only a slight provocation (Steffen et al. 2016, p. 326; recently, the Leopoldina, the German National Academy of Sciences, has made the case for a deeper integration of Earth system sciences into the Geosciences, cf. Deutsche Akademie der Naturforscher Leopoldina e.V. 2022). Secondly, and this is much more important with regard to the provocative potential, stratigraphy has never before operated with a term that puts a specific actor in its center—even more so a species, that, with regard to the history of planet earth, is nearly as briefly a factor as the Anthropocene (see e.g., van der Pluijm 2018, 2014). Traditionally, the nomenclatural rules, fixed in the ISG, require that the name of a geochronological time unit refers to the geographical location of the defining GSSP. Admittedly, stratigraphy no longer meets this set of rules in the definition of geochronological units at the level of epochs since the beginning of the twenty-first century: the terms in question (Paleocene to Holocene) do not indicate a geographical reference. Instead, the first part of the determinative compound functions as the compound's modifier of the compound *cene*. Nevertheless, all of these epoch labels are neutral descriptions of states. By conceptually anchoring a species on the level of epochs, stratigraphy not only leaves the stratigraphic set of rules behind, but furthermore abandons the neutral ground and catapults itself—perceived by its representatives as objective and value-free—right into the pool of basic anthropological questions. The classical dichotomy of "observed object" and (human) observer that counts for every other unit of the GTS cannot be maintained any longer and represents a deep epistemological provocation. The debate continues, and one can only speculate whether the decision would have been different if Paul Crutzen had proposed something like a Crawfordocene.

But bringing the *anthropos* into play in a discipline whose expertise focuses exclusively on the analysis of rocks and sediments is not only to be understood as a provocation but also as an invitation (Interview Zalasiewicz, 2017). Indeed, the consideration of humans and their actions in geological debates is not completely new. After all, the definition of the Holocene is, besides other aspects, based on phenomena related to human beings, such as the development of fire use becoming the most important cultural technique as a result of settlement and domestication during the Neolithic Revolution (Zalasiewicz et al. 2008). What is new about the Anthropocene, however, is the dimension of human impact. Whereas "[h]uman activity […] may help characterize Holocene strata, […] it did not create new, global environmental conditions that could translate into a fundamentally different stratigraphic signal" (Zalasiewicz et al. 2008, p. 5). It is the latter that distinguishes the anthropocene human activity from its holocene counterpart: In the Anthropocene, humanity is no longer just one of many factors that contributed to a fundamental change in the Earth system, but the central, "pre-dominant" factor (Gibbard 2008). The Anthropocene consequently urges stratigraphers to take into account topics that have traditionally been within the realm of humanities and social sciences, such as questions regarding human responsibility for planet earth. This call comes along with the challenge to step out of established disciplinary expertise and comfort zones and enter an active dialog with representatives of other disciplines. It is obvious that the term Anthropocene exerts an almost magnetic attraction on all those disciplines that deal with questions concerning humanity.

Internal geological reactions to methodological innovations, experimental approaches and changes with respect to the stratigraphic evidence production process—with the *Anthropocene Working Group* as the personification of this novelty—have often been characterized by criticism and opposition in the course of the two decades of study of the Anthropocene. As recent developments, namely the rejection of the Anthropocene proposal by the SQS, show again, insight and demand for methodological adjustements based on the requirements of the most recent stratigraphic deposits didn't gain acceptance in the broader stratigraphic community (Finney and Edwards 2016; Walker et al. 2018; Bauer and Ellis 2018; Lewis and Maslin 2015a; van der Pluijm 2014; regarding the provocative potential for geology see, for example, Visconti 2014).

Initiated by Paul Crutzen in 2000, but most recently with the establishment of the AWG in 2009, stratigraphy moved into an Anthropocene spotlight that is not only unfamiliar to the actors involved but also calls them to leave their own disciplinary arena in favor of entering an active inter- and trans-disciplinary exchange. Closely related to this is the concern that the quest to ratify the Anthropocene as a new geological time unit has been seen as a political as much as a scientific endeavor (Finney and Edwards 2016; Autin and Holbrook 2012; Visconti 2014; Baskin 2015; Edwards et al. 2022).

The normativity of the concept, as well as the unusually high media and interdisciplinary interest in this specific geological decision-making process, exposes the transformative character of the Anthropocene in its multidimensionality in a particular way. Therefore, it seems hardly surprising that traditional, established processes of geological knowledge production are facing a crisis. As the following will show, methodological, structural, and content-related boundaries get blurred and trigger negotiation processes in novel constellations. This becomes especially evident within the AWG, but also in the debate about the Anthropocene in general (Will 2021, pp. 105ff.).

Chapter 2 has already explained the composition of the AWG. In contrast to the other geological working groups, as they exist for each epoch of the Geological Timescale, the AWG consists of members from different disciplines. It is exactly this interdisciplinarity that makes the working group unique within geology.

Members of stratigraphic working groups are not elected. Instead, it is the task of the appointed Chair to form them on their own and to ensure their functionality. Jan Zalasiewicz was convinced from the beginning that the Anthropocene could only be approached in an interdisciplinary way, also from a stratigraphic perspective. Early on, after being entrusted with the formation of the AWG in 2009, he therefore formulated the goal of involving archaeologists and historians in particular, representatives of the two disciplines that, in addition to stratigraphy, deal quite explicitly with temporalities, "to cover fields of research both in 'deep time' and contemporary environmental change" (Zalasiewicz 2009). Furthermore, it seemed essential to include Earth system scientists in the working group—especially since the concept originated in the context of Earth system science: "[I]t had to go beyond the usual circle" (Interview Zalasiewicz, 2017). Zalasiewicz was aware that he was taking quite a risk with the planned novel composition of the working group: "[I]n a way it was a high-risk strategy because we could have produced something that would've been dysfunctional" (Interview Zalasiewicz, 2017). Undoubtedly, the internal working process was a learning process for all AWG members. The Anthropocene Working Group had to prove, not only to the outside world but also to itself, that it was geologically functional in its interdisciplinarity. Recent developments in the AWG work process show that it has succeeded in doing so.

The discussions surrounding the composition of the AWG as a stratigraphic working group, which pursues the goal of addressing a geological question by means of stratigraphic analysis, reached far beyond the inner geological circle and stimulated tense discussions about responsibilities and legitimation sovereignties. This was at least as novel as the interdisciplinary composition of the group itself. Despite persistent criticism on the lack of judgment of representatives of external disciplines on geological issues, the AWG has not deviated from its course and until 2020 has involved all its members in the corresponding votes. In 2020, the AWG decided to distinguish between voting and advisory members for the process of GSSP selection in order to avoid potential criticism of the formalization proposal based on alleged methodological inaccuracies. This development does not represent a disciplinary takeover, but is rather of a strategic nature. As the latest developments have shown, namely the rejection of the formalization proposal by the SQS in March 2024, these strategic endeavors didn't work out for the time being. But this doesn't mean that the Anthropocene will disappear. It is here to stay. Humanities discussions about the concept will continue and are perhaps more important than ever before—not least because they are in some ways freer than geology, as they do not require a common definition arrived at by a vote of a few people. We can be sure that the geological debate about the Anthropocene will also continue. As things stand, the AWG will continue to exist as an independent body despite its exclusion from the ICS. A new formalization process can be initiated and does not need to wait for 10 years (cf. preface). Nevertheless, for now, the impression remains that the SQS's decision is partly based more on sensitivities and methodological disputes than on the evidence itself: "Many geologists, used to working with millions of years, find it hard to accept an epoch just seven decades long – that's just one human lifetime" (Turner et al. 2024).

Based on the 25 expert interviews that I've conducted with eight members of the AWG and 17 key representatives of the debate from the humanities, social, and natural sciences, the self-perception and external perception of the AWG in the context of the Anthropocene debate have been surveyed. While three of the 17 actors from outside the group consider the geological discussions around the concept irrelevant and are indifferent toward the question of formalization, 14 of the 17 respondents consider a ratification desirable and perceive the geological debates as a stimulating impetus for a far-reaching debate about the past, present, and future of human impact on planet Earth. None of the interviewees from outside the group critizes or even debates the interdisciplinary and in geological terms "contested" nature of the AWG. On the contrary, all of them highly value the interdisciplinary composition of the working group. Furthermore, based on the geological debate and the involvement of humanities scholars and social scientists in the stratigraphic research process, they see great potential in the concept to initiate new collaborations beyond disciplinary boundaries. Interestingly, all 17 of the interviewees from outside the group consider the formalization of the Anthropocene to be outdated—they all hold the view that the Anthropocene has long been recognized outside geological circles: "I actually think what they do is irrelevant to people in the humanities, in the arts. The genie is out of the bottle, everybody is using it. […] [T]he popular decision has gone ahead of the professionals" (Interview Robin, 2017). At the same time, the early opening of the AWG to perspectives from the humanities and social sciences is considered to be of major importance. This alone, and the early media echo associated with that (see Chap. 5), paved the way for a broad discourse on the concept in the humanities and in the public sphere. The only point of criticism formulated by three interviewees relates to the possible implications that a decision for a birthdate of the Anthropocene would come along with: "What I think is perhaps not being done as well as that is considering the output, the implications of what their decision may be and how that will reverberate back out to society" (Interview A2, 2018).

Despite the mostly positive perception of the AWG from outside, there was repeated indication from the interviews with non-members that there exists indeed a certain feeling of marginalization toward the geological debate among scholars from the humanities and social sciences. Repeated argumentation against the geological approach, which is perceived as presumption and encroachment on definitional sovereignties, gives voice to this feeling (see, for example, the debate on twitter provoked by Erle Ellis' publicly posted resignation from the AWG; Ellis 2023a, b). Interestingly, in this regard, interviews showed a somewhat opposite tendency among the AWG members. There exists, for instance, no desire to reduce the various perspectives from the humanities, social sciences, and natural sciences to one common definition. The aim was rather to stimulate discussion, whereby the definition of what the Anthropocene is needs by no means be identical across disciplinary boundaries (Interviews Waters, 2017; Steffen, 2018; Zalasiewicz, 2017; A4*, 2017; A3*, 2017; Williams, 2017; McNeill, 2017; in more detail, see Will 2021, pp. 105ff.). It is true that the AWG was concerned with launching a formalization proposal for the Anthropocene to add the Anthropocene to the GTS as the most recent geologic epoch. For this, the AWG needed to present a solid stratigraphic analysis that clearly defines what the Anthropocene means in a geologic sense. This by no means implies that all other disciplines must align themselves exclusively with this geologic definition of the Anthropocene. Rather, it was intended to serve as a common reference point for a concept whose scope extends beyond a physically existing geological epoch (Zalasiewicz et al. 2021, p. 9). John McNeill puts it as follows: "It seems to me that there is no fundamental incompatibility between what the stratigraphers wish to do, and what humanists wish to do. But I am interested in whether or not there is sufficient logic to understand the Anthropocene as a period of human history in addition to understanding it as an interval in Earth history. And it's quite possible they should be recognized as both. It's quite possible they should be recognized as one and not the other. And if it is possible that they should be recognized as both, it's not necessarily the case, they should be identical" (Interview McNeill, 2017).

However, the debate about the Anthropocene as a cultural concept might make an important contribution to future-oriented questions about the (intergenerational) responsibility of humans for the planet, as the following quote from Will Steffen illustrates: "So I think what this does about the future is it puts some real fundamental questions out for the humanity scholars, the social science scholars, the governance scholars, you know, the fundamental questions like how do you define human progress how do you actually define that now. And that's not a natural science question" (Interview Steffen, 2018; cf. also Rockström et al. 2023). This is consistent with Sheila Jasanoff's assessment "that, despite significant changes in the resources with which we can rethink the human condition, drawing upon the sciences, history's fundamental purposes have not been rendered irrelevant. [...] In particular, the question of how humans should imagine the stewardship of the Earth in the Anthropocene remains an ethical project for history and not primarily the domain of the natural sciences" (Jasanoff 2020).

4.2.2 The Anthropocene as a Provocation to Humanities and Social Sciences

The debate about the Anthropocene as a cultural concept begins while the geological debate about the age of humans is already well underway. The discussions in the humanities and social sciences about various aspects of the Anthropocene concept further develop especially from 2013/2014 onward, as can be seen from the exponentially increasing number of publications by representatives of the humanities and social sciences (see Brauch 2021). The launch of the three interdisciplinary Anthropocene journals, in particular the journal *The Anthropocene Review*, contributes to a stronger focus on humanities and social science aspects of the concept, and the growing fields of discussion around Anthropocene issues in the following years demonstrate the enormous analytical and heuristic potential of the concept (on the increase in humanities literature around 2013/14, cf. also Castree 2014). The Anthropocene concept, which in terms of terminology puts the human species at the center of attention, "repositions humankind in the universe, repositions humankind on the earth, repositions our understanding of what is nature and what is not nature" and thus challenges established core assumptions (Interview McNeill, 2017).

The provocation for the humanities and social sciences, which results from the thesis of the geological impact of humans, is consequently at least as significant. But, in contrast to the geological provocation, it emanates from different aspects. While geology is challenged primarily in terms of methodology, the humanities and social sciences are challenged to rethink established conceptual frameworks. Well-established narratives that are based on linear temporalities and operate within hitherto clearly distinguishable categorical systems seem to be no longer sufficient in the face of anthropocene questions.

What provokes most obviously is the term itself. Who is the *anthropos* in the Anthropocene? The use of the species concept as an analytical category, which is not at all problematic in geology, triggers highly tense discussions about the question who constitutes this "we" in the humanities. Moreover, despite the postulated irrelevance of ethical aspects, the understanding of humanity as a central actor in a planetary perspective is diametrically opposed to the problematizing and differentiating perspective of the humanities and social sciences—this points again to the question of iso- or diachronous change. Also, humanities and social sciences often perceive the orientation of the natural sciences, especially geology, toward the anthropogenic domain as a threat to their definitional sovereignty. By examining the Anthropocene, natural scientists are entering the pool of fundamental anthropological questions—a field of research previously reserved for disciplines from the humanities and social sciences. Historians in particular, who are generally regarded as experts for the periodization of human history, consider themselves particularly challenged by the geological examination of the Anthropocene as a new geochronological time unit. Jan Zalasiewicz sums up this conflictual struggle for interpretive sovereignty with respect to the category of human time by telling about an encounter with a historian who "accused me and another geologist of stealing time" (Interview Zalasiewicz, 2017). This conflictual struggle is still present at the time of writing as the discussions in the wake of the GSSP announcement show (Ellis 2023a). However, there are also examples that point to a genuine co-thinking and demonstrate real efforts to bring together the different disciplinary perspectives in a fruitful way. A book published in 2022 by Jan Zalasiewicz and the historian Julia Adeney Thomas is one of those examples (Thomas 2022a). Thomas makes a plea for telling new stories for the new reality of the Anthropocene. Instead of doing this in the mode of "Anything goes," which is unable to overcome the division between human history and earth history, she calls for developing new narratives that start from the geological evidence via the modes of the "Singular Story" and "Democracy of Voices," which take into account the global dimension as well as different local realities and are based on a common understanding of the future challenges (Thomas 2022b).

The tense provocative potential of the debate about the Anthropocene as a cultural concept manifests itself partly in broadening discussions of existing debates such as the climate change or sustainability debate, but partly also in the development of new lines of discussion. With the planetary dimension of anthropogenic action, which is the overarching starting and ending point of any anthropocene discussion, it is, however, in any case discussions under altered conditions. It is precisely this planetary dimension that calls into question the usefulness of established disciplinary research practices and narratives. Instead of narrating processes starting from local phenomena, the Anthropocene, in the spirit of a reversal of the bottom-up thinking, calls for localizing, differentiating, and specifying the planetary, the collective, and the abstract. The accompanying provocations in terms of content attest, for example, to the frequently voiced criticism that the Anthropocene was a Western and eurocentric concept. Moreover, it was by far too anthropocentric and in an ethical perspective untenable—since, by treating the *anthropos* as a collective entity, it would burden every inhabitant of the earth with the same amount of responsibility for the current situation of the planet. By treating these provocations as an opportunity, the humanities and social sciences have the chance to take on the urgently needed role of a mediator and translator between the planetary and local dimensions of the subject (Demos 2017; Kamel 2022; LeCain 2015).[3] For there is hardly anything our society needs more than a multilateral dialogue about human responsibility for planet earth in the past, present, and future that overcomes traditional boundaries.

The key methodological difference as well as the primary cause of misunderstandings and disputes at the interface of the disciplines involved originates in the effect-centeredness of stratigraphy, which is diametrically opposite to the cause- or process-orientation of disciplines in the humanities and social sciences. Both perspectives are crucial and may complement each other, but they also need to be considered separately:

> We stress the importance of differentiating between the long *process* leading to the Anthropocene, from the Late Quaternary Extinction Episode to the start of the Industrial Revolution, from the results of the Great Acceleration, leading to the geologically sudden onset of a different Earth System with a different sedimentary expression. Disentangling the process from the result is important, especially to help understand how the Anthropocene became reality (Waters et al. 2022, 22).

In the intra- and interdisciplinary arena, empirical facts and contexts get interlinked and thus transform into narratives with different emphases. In this regard, anthropocene topics prove to be particularly challenging insofar as established narratives, often characterized by linearity, prove to be no longer sufficient for these new stories. Admittedly, the narrative overload is sometimes triggered by the planetary scale. But the underlying causes are more complex. The criticism that the Anthropocene is an anthropocentric concept comes first and foremost from the humanities and social sciences. Consequently, the respective disciplines have started to explore the agency of nonhuman actors—be it in a material or technological perspective—and to imagine Anthropocene futures, with or without humans as the inhabitants of the planet.

Each of these perspectives provides a different focus, and it has become common practice to frame that interest under an alternative, or rather specifying, term.

The following figure maps the terms introduced into the debate from 2000 to 2024. They distinguish themselves by putting different actors, processes, or substances in the center with regard to the new geological epoch. In addition, the terms differ in their temporal perspective. While proposals such as the *Machinocene* by the philosopher Huw Price or the *Robotocene* by Roger Søraas and Håkon Fyhn (Science and Technology Studies) point to future-oriented conceptions of technology, Tim LeCain's *Carbocene*, Christoph Bonneuil's and Jean-Baptiste Fressoz' *Anglocene*, or Steve Pyne's *Pyrocene* refer in a more cause-oriented perspective to processes and actors in the past that have paved the way into the Anthropocene.[4] In the end, the alternative designations for the age of humans all reflect either "a

[3] On these criticisms of the Anthropocene concept, cf., for example, Manemann (2014), Yusoff (2018).

[4] The four scholars mentioned, Tim LeCain, Christoph Bonneuil, Jean-Baptiste Fresooz, and Steve Pyne, are historians.

chosen part of the set of diagnostic characters or [are] providing a suggested explanation for the causes of the epoch's existence" (Zalasiewicz et al. 2019a, p. 15).

More than 100 alternative terms collected sofar show that the Anthropocene has become a concept with multiple meanings (Chwałczyk 2020; Will 2021, p. 208). From a metaperspective, the different perspectives reflected in the terms are not, as is often perceived, in an oppositional relationship to one another.[5] Rather, the proposals complement each other in terms of their content. The number of alternatives, with some certainly to be taken more seriously than others, at the same time points to the fact that the phenomena described by the Anthropocene are accepted and not called into question anymore. The terminological provocation has proven to be a fruitful starting point for the negotiation process about the Anthropocene phenomenon in the humanities and social sciences (Fig. 4.1).

By looking at the content-related characteristics of the individual terms, the range of disciplines involved as well as the multidimensionality of the epistemological interest are clearly shown.

As sociologist Bronislaw Szerszynski aptly puts it, however, it is not a matter of identifying a hierarchically superordinate Anthropocene narrative that overrides all other narratives that are part of the debate. Rather, the Anthropocene is to be seen as a metanarrative in a planetary perspective that combines many different meanings, all of which have their own validity:

> [Y]ou could say the Anthropocene is itself an example of the Anthropocene story, which is a kind of a product of the technosphere. It's a sort of abstract machine that moves around the earth and incorporates more and more local narratives into this big global meta-narrative. So it's sort of exemplifying in a way, the Anthropocene narrative is part of the Anthropocene machine, so without completely abandoning all the new understanding of the planet. This is some way we can retell that story, which in a way which is less European in feel really, it has more how things look from other less dominant cultures (Interview Szerszynski, 2018).

The Anthropocene is a concept with multiple meanings. Many of them vary both in their emphases and in their disciplinary selectivity, but at the same time complement each other.

The impossibility of defining distinct areas of responsibility at the level of content leads to a blurring of boundaries, both at the disciplinary level and at the level of content. At the same time, this provides the humanities and social sciences with the chance to enter into a symbiotic relationship with the natural sciences in the spirit of cross-disciplinary cooperation. "I think we're at the moment where [...] [the] cultural divide has begun to collapse" (Interview LeCain, 2018). No less than the relationship between established dichotomies such as nature and culture, subject and object, environment and society are at stake in the debate about the Anthropocene.

[5] The Capitalocene is an exception here insofar as it has triggered a genuine debate about responsibilities and tense discussions between representatives of different disciplines. Nevertheless, from a causal perspective, methodological differences are in the foreground (effect vs. process orientation).

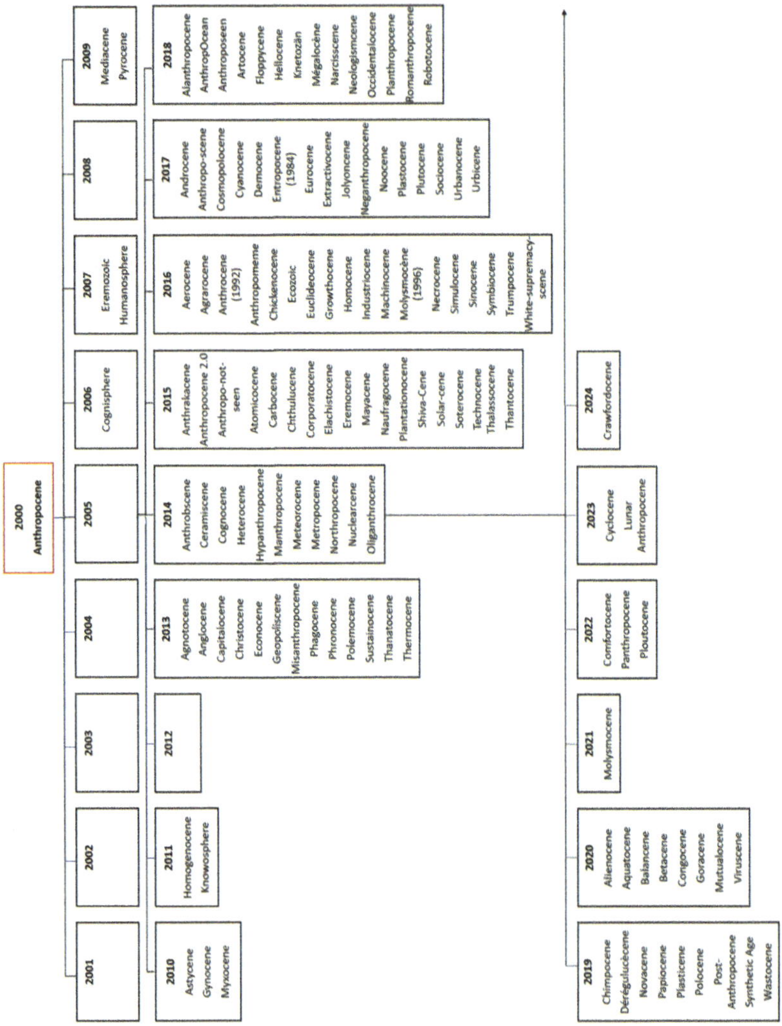

Fig. 4.1 Alternative terms for the Anthropocene from 2000 to 2024

4.3 Challenged Interaction: About the Promising Effects of Negotiation Processes Beyond Disciplinary Boundaries

However, interdisciplinary cooperation turns out to be fraught with tension in the debate on the Anthropocene as well—and it is only productive when the different perspectives stand side by side on equal footing and not in competition with each other, since it is only this multiperspectivity that takes account of the complexity of the concept. Despite various promising examples, there is still a long way to go here. One of the conceptual challenges to be addressed in this context is to bring together the different temporal and spatial scales that merge in the Anthropocene. "[W]e are equipped to live in very local contexts [...] at a local environment [...] we are not existentially or mentally equipped to deal with a planet with 7 billion people for 100.000 years. It's just beyond our capacity" (Interview Hornborg, 2018).

Cooperation beyond disciplinary boundaries has the potential to overcome these (mental) barriers and to advance not only interdisciplinarity within academia—a step that is not only necessary but even overdue in the face of the *wicked problems*, global, complex challenges that offer no clear solution, that we discuss in reference to the Anthropocene. Furthermore, such a cooperation can generate synergies and result in the development of new lines of thought that can also advance communication toward the non-academic realm.

In the following, this potential will be exemplarily explored by means of two categorical systems that characterize modern thinking and have become unsettled in the Anthropocene: the concepts of time and space and the subject-object dualism. Similar to the diagnosis that established concepts of space and time of the Western tradition of thought are no longer sufficient in the Anthropocene, an expansion of the spectrum of actors is emerging with the latter, which catapults natural sciences into the pool of basic anthropological questions, while humanities and social sciences are required to overcome the nature-culture dichotomy and to integrate more-than-human-agency into their considerations.

4.3.1 Time as a Contested Category

With the Anthropocene, humanities face a challenge that consists primarily in the integration of the the planetary, be it with regard to deep-time or spatial scales (for an overview, see Antweiler 2022). While geology is challenged to combine small with large, the reverse is true for the humanities and social sciences. Geologists need what humanities scholars have, humanities scholars lack what geoscientists deal with on a daily basis. This offers great potential for mutual enrichment, the exchange of expertise, and complementary knowledge production, which can be achieved by means of an open, active negotiation process at the interstice of disciplines.

Historians, as experts for the history of mankind, are directly affected by the discussions about the Anthropocene (Trischler and Will 2019; Westermann and Höhler 2020; Quenet 2017). It is no coincidence that the Anthropocene is being discussed particularly intensely in the history of technology and environmental history. Whereas the temporal fields of responsibility between geology and history have so far been clearly separated from each other, for the latter, nothing less than the defining power over the category of human time is at stake with the geological appropriation of the human timescale, which history has claimed as *its* time since the differentiation of the disciplines in the nineteenth century. For centuries, the humanities and social sciences have been treated as soft or secondary sciences that have to permanently prove not only their legitimacy but also their usefulness. Mainly from the side of the natural sciences, they are sometimes perceived as having an auxiliary scientific character (Snow 1967; Vowinckel 1994; Cahn 1991; Mittelstraß 1991; Daston 1998; Halfmann and Rohbeck 2007). It is against this background that the anthropocene trial of strength is to be examined. It manifests itself partly in openly expressed competition about the definitional sovereignty.

Since the Anthropocene thus reopens old wounds, the accusation of a marginalization of the humanities and social sciences in the Anthropocene debate by the natural sciences, which are addressing subjects that traditionally are not part of their field of expertise, is understandable (Castree 2017; Sörlin 2018b, p. 22). However, such an objection fails to recognize that the broader natural science community in general and the AWG in particular are exclusively concerned with the geological division of time: "we're not trying to periodize human time at all, we are working on understanding Earth time […] it's a nonexclusive interpretation" (Interview Zalasiewicz, 2017).

Obviously, the perception of being marginalized does not apply entirely to the representatives of the humanities and social sciences involved in the debate. Many of them take a consistently reflective stance, as the following quote from an interview illustrates: "[H]istorians don't say to themselves, 'What do you think of that rock strata?'" (Interview B1, 2018). On the contrary, historians raise entirely different questions about the Anthropocene and, in contrast to natural science disciplines, take a cause- and process-oriented view. This perspective explains why the practice of dividing time in historical terms is based on different periodizations than the geological one. However, these are not necessarily mutually exclusive. Although stratigraphy repeatedly refers explicitly to the effect-centeredness of geological practice, it emphasizes that it considers an orientation toward the underlying processes to be important as well. To be sure, this should remain within the competence of the humanities. However, dialog is welcome, as the following quotation impressively illustrates: "I'm very glad the humanities are taking the interest they're doing […] because […] if you are to understand the Anthropocene, you have to understand the driver. We're rock people, we are not very good with humans" (Interview Zalasiewicz, 2017).

Both the interdisciplinary composition of the working group and the role of the humanities scholars and social scientists within the group demonstrate that an orientation toward processes is also significant for the stratigraphic Anthropocene. The work of the Anthropocene Working Group has shown in practice that the humanities approach to the Anthropocene, also to the stratigraphic Anthropocene, is essential—after all, the beginning of the geological Anthropocene, which is located in a historically incisive period, cannot be understood without its contextualization. Colin Waters and Simon Turner, in the press conference announcing the GSSP on July 11, 2023, once again stressed the historical and societal significance of the 1950s alongside the geologically traceable fundamental change in the GSSP selected by the AWG for the Anthropocene as an advantage for the broader debate and future implications at the level of understanding and action (Klingan et al. 2023). The Anthropocene is here to stay and so is the broader debate about the concept outside geology, which now with the rejection of the formalization proposal by the SQS will possibly become even more important than in the past years.

Nevertheless, the blurring of boundaries is in parts perceived as a threat—so to speak as a destabilizing-opening moment that needs to be restabilized and closed at any cost. However, there is an opportunity here to use the destabilizing moment in the sense of interdisciplinary cooperation between the humanities and the natural sciences and to systematically advocate for the relevance of one's own discipline: "[H]umanistic disciplines at least perceive themselves to be in crisis, they struggle to be judged as relevant and useful. And so saying things about the Anthropocene is a way to compete for relevance, for recognition, for legitimacy" (Interview McNeill, 2017). This can be achieved, for example, by emphasizing the fact that scientific meaning-making and explanation for the Anthropocene are impossible without reference to knowledge from the humanities. The establishment of new hybrid disciplines such as the *environmental humanities*, the *energy humanities*, or the *integrative humanities* within the last two decades in response to the environmental issues widely discussed in the climate change and sustainability debates, which are also part of the debate on the Anthropocene, can be seen as another way to emphasize the significance of humanities knowledge with respect to the production of natural science knowledge (Sörlin and Wynn 2016; Sörlin, 2018a, b; Palsson et al. 2013; Gauger and Rüther 2007).

The interviewees from the humanities and social sciences agree that the Anthropocene—if it is to be used as a concept in all its facets—can only be approached on the basis of altered knowledge production mechanisms (Interviews B1, 2018; B3, 2018; B8, 2018; Robin 2017; Szerszynski 2018; LeCain 2018; Pádua 2017; B2, 2018; Hornborg 2018; B5, 2018).

Not all of them go as far as Dipesh Chakrabarty, who in his paper *The Climate of History: Four Theses* published in 2009 calls for a climate change in the historical sciences, away from the distinction between the separate categories of nature and culture or environment and society, toward an integrated geo-history (Chakrabarty 2009b, 2015, 2016, 2021). Known as a thought leader of postcolonial studies, he has an influence within the historical community that the reception of and engagement with his theses reflect (Chakrabarty 2009a). A study compiled by Robert Emmet and Thomas Lekan is exemplary for the multiperspective discussions of the relationship between humans and the environment that he initiated (Emmett and Lekan 2016; cf. also Trischler and Will 2019).

One of the early projects working across disciplinary boundaries is the initiative *Integrated History and future Of People on Earth* (IHOPE). It emerged from the IGBP and was established as an independent project of the IGBP at a conference in Dahlem in 2005 (Costanza et al. 2012; Costanza et al. 2007a; Costanza et al. 2007b). The purpose of the initiative is to understand the interaction between humans and the environment in a historical perspective over a period of 100,000 years. Therefore, the Global Change Community responsible for the project has set itself the task of providing an overview "of how (and why) the planet and human societies have changed in historical times. [...] Initial development of IHOPE histories will begin with local and regional scale interactions between humans and their environment and lead up to global scale issues" (International Geosphere-Biosphere Program 2010).

As a historical project that was not initiated by the historical community and, moreover, has been carried out by an interdisciplinary group consisting of archaeologists, geographers, historians, environmental scientists, earth systems scientists, computer scientists, anthropologists, and biologists, it both provokes and stimulates scholars involved. Historian Libby Robin, who came into contact with IHOPE immediately after its inception through earth systems scientist Will Steffen, and has been engaged in anthropocene content ever since, is an excellent example. In the essay *History for the Anthropocene*, coauthored with Will Steffen, she writes, "the 'Global Change community' challenges historians to be interdisciplinary and beyond 'normal' scales of history, both in time and in space" (Robin and Steffen 2007, p. 1704). A figure published by Robert Costanza et al. (2007b, p. 524) illustrates this point.

In this figure, the members of the IHOPE have dared to carry out the frequently demanded merging of geological temporality on the one hand and historical temporality on the other, not only in theory but also in practice (as a model). Vertical geological time and horizontal historical time converge in the graph, which allows the recipient to relate increases and decreases in individual scientific parameters, such as methane or CO_2 concentrations in the atmosphere, to specific historical processes and events—at least from a temporal perspective. This is because the authors explicitly point out that this first effort to bring together different scales does not yet depict a causal relationship. While political and cultural events are located on the left side, the right side shows technological developments. A closer look reveals that the listed events are often related to environmental-historical processes as well. Moreover, the geological "time-space" of 100,000 years loses its abstraction when combined with aspects of historical time.

A second, very recent example, is from a paper published by members of the AWG in *Episodes*, the most important geological journal. In a figure that is part of the paper they combine geological and historical timelines "using empirical data to reveal the true scale and timing of social and environmental changes" (Head et al.

2023, p.234) and to visualize the interconnectedness between environmental and social changes.[6]

With this new kind of research at the interface of natural sciences and humanities, IHOPE as well as the approach by members of the AWG is to be understood as an invitation and stimulus for historians—as an invitation to take up the challenge of the Anthropocene and as a stimulus to conceptualize layered models of historical temporalities.

Robin herself has taken on this task. Her continuing engagement in the debate to the present day demonstrates this (IHOPE has resulted, among others, in Robin et al. 2013). In a 2015 paper, for example, she argues for a common language across disciplines that allows us to unite "the numbers of natural science and the metaphors of humanity" (Robin 2015, p. 20). In 2017, in collaboration with the historians Paul Warde and Sverker Sörlin, she takes up the Renaissance comparison that has been formulated by Stanley Finney and Lucy Edwards, among others, as a criticism of the geological Anthropocene concept (Finney and Edwards 2016, p. 8f.; Finney and Gibbard 2023). Both the Renaissance and the Anthropocene are concepts with multiple meanings. For both concepts, depending on the point of interest, different periodizations are applied (Warde et al. 2017). At the same time, however, the criticism put forward by Finney and Edwards turns out to be misguided insofar as they use an aspect from the field of expertise in the humanities, and even more from the practice of periodization in the humanities, as the basis of their geological argument. To a certain extent, they thereby trap themselves by practicing what they criticize the AWG for: They argue in an interdisciplinary way by mobilizing knowledge from outside the discipline as a resource to strengthen their own thesis. The same can be seen recently in the proposal to treat the Anthropocene as an event, but not as an epoch: Here, too, interdisciplinary applicability serves as an argument for a geological thesis (Gibbard et al. 2022; for details in this respect cf. Westermann and Höhler 2020).

Global historian John McNeill has been involved in the Anthropocene debate from an early stage. His involvement is not limited to his AWG membership and the geoscientific debate (Steffen et al. 2007; Zalasiewicz and Waters 2012). The humanities debate also benefits from his publications and lectures on the Anthropocene, which differs from other historians' engagement with the Anthropocene because of his direct access to and knowledge of the origins of the debate and the stratigraphic discussion. He is thus given a contextualizing and explicative function in two respects. While he is responsible within the Anthropocene Working Group for the historically correct contextualization of natural science knowledge, one may ascribe the same function to him for humanities discourses, in which his insider perspective allows him to act as a geological corrective (cf., for example, McNeill 2019; Interview McNeill 2017; Interview Zalasiewicz 2017).

In his book *The Great Acceleration*, coauthored with Peter Engelke and published in 2014, he tells a global environmental history of the Great Acceleration

[6] https://www.episodes.org/journal/view.html?doi=10.18814/epiiugs/2022/022025.

from the end of the Second World War onward and thus makes an important contribution to understanding the sociopolitical background of the new geological epoch. Thereby, he not only builds upon previous historical studies such as those of Christian Pfister or Rolf Peter Sieferle, whose work already pointed to a fundamental change around the middle of the twentieth century, but at the same time establishes a link to the *Great Acceleration Graphs* published in 2004 as well as the discussions within the AWG about the periodization of the new geological epoch (Pfister 2010; Sieferle 1982, 1994). In doing so, McNeill and Engelke succeed in establishing references to geological subdiscourses. For example, they take up the marker question and argue that the bones and teeth of mammals born in the 1940s and 1950s should be examined for their level of radioactivity. The latter should be significantly different from what can be detected before 1940 and after 1963 (McNeill and Engelke 2014, p. 167). In the meantime, this hypothesis has been confirmed: The bones of broiler chickens, which, as biofossils, serve as one of the secondary markers of the Anthropocene, show significantly elevated levels of nitrate and carbon isotope valules compared to the levels before 1950 (Bennett et al. 2018). With this, McNeill and Engelke also allude to a relatively new research field for the historical sciences that belongs to the field of deep history: aDNA research (ancient DNA) (see overviews by Bösl 2017a, b). This field evolved in the 1980s out of a collaboration between the natural sciences and the humanities and today offers a fruitful opportunity for cooperation across disciplinary boundaries for the historical sciences as well as other disciplines in the humanities and social sciences (for historical projects that use the aDNA method for their research or consider it to be useful, cf., for example, Geary and Veeramah 2016; Samida and Feuchter 2016; Thomas 2014). McNeill and Engelke's book also successfully draws attention to fundamental anthropological questions, such as human responsibility for the planet, and argues for the relevance of humanities, especially historical, engagement in the debate. The latter, McNeill notes, is far from exhausted (Interview McNeill 2017).

This may be true for the historical sciences in general—but not for environmental history in particular, which, beyond the temporal aspect, is particularly affected by any aspect of the Anthropocene debate. It is especially environmental history that can handle the challenges of an Anthropocene historiography more easily than historians with a different focus. After all, environmental history frequently moves "audaciously across time and space and species and thereby [anyway] challenges some of the conventions of history" (Griffiths 2002, p. 377).

There is now a phlethora of environmental history contributions and works from the flourishing field of environmental humanities (cf. exemplarily Oppermann and Iovino 2017; Castree 2014; Wilke and Johnstone 2017; Horn and Bergthaller 2020). Early works from environmental history include Paul Dukes' *Minutes to Midnight*, in which he tells the story of the age of humans starting from the end of the Seven Years' War in 1763 through the accelerated scientific and technological progress that began with industrialization. He not only acknowledges the remarkable merit of the Anthropocene, namely linking geological and historical time, but further advocates a pandisciplinary approach. Gregory Cushman's *Guano and the Opening of the Pacific World* also joins into the line of approaching the concept from an environmental history perspective (Dukes 2011; Cushman 2013). Following the posthumanist approach of considering nonhuman entities as actors, Cushman focuses on guano, the fertilizer with high content of phosphorus and nitrogen, which was primarily extracted in South America and then exported from the mid-nineteenth century onward. He highlights the global historical consequences of the widespread use of this substance and thus implicitly argues for a historical beginning of the Anthropocene around 1850.

Bronislaw Szerszynski thinks of the Anthropocene as a kind of comprehensive global environmental history: "[T]hat became a sort of global environmental story. [...] [A]nd for me most of the time I kind of think it is a pretty amazingly powerful story" (Interview Szerszynski 2018).

In their *History Manifesto*, published in 2014, Jo Guldi and David Armitage argue for curing the short-term thinking that characterizes contemporary societies with the concept of the Anthropocene. According to them, the concept can be used as a tool to trace back society and politics to the *longue durée* by enabling historiography to look at past phenomena that, taken together, have led to the situation in which humanity finds itself today (Armitage and Guldi 2014). For their plea for a shift back to long-term thinking, they have been widely praised. However, with the intended turn away from *short termism* and toward *long termism*, Guldi and Armitage remain within the established dualistic category system that representatives of *deep history* have already left behind. Besides that, the manifesto also provoked criticism (Cohen and Mandler 2015). Quoting sociologist Joshua Yates, they call "[to] realize our fullest potential as managers of the earth and our future on it" (Yates 2012, p. 22). Analogies with the belief of ecomodernist thinkers in the *technological fix*, which the latter promote in the *Ecomodernist Manifesto*, cannot be denied (Asafu-Adjaye et al. 2015).

The normative charge of discussions in the humanities due to the link to humans as the object of research is always present. Since it is part of the task of the historical sciences in particular to position themselves normatively—the aim, after all, is to gain knowledge from the past and to derive future action from it—it is rather content-related aspects beyond normative questions that affect history in a provocative way.

Humanities and social sciences scholars often take up specific aspects of the concept that fit into their own field of interest and transform and interpret them according to their own disciplinary rules (for an overview cf. Will 2021, pp. 209ff.). Since there is now a consensus that there is not the one Anthropocene, but that the formerly earth-system science idea has differentiated into a concept with multiple meanings, the way in which historians engage with the concept has the potential to be complementary with respect to the general debate, but also with respect to the geological debate. Examining the work of historians involved in the debate has revealed that the central challenge to the discipline of history is to bring and think together geological, historical, and human temporality:

> Usually we deal with decades, we're talking about long 1970s, and the short 1980s, and have continuous debates about such a small-scale period. We are not trained to look at this long view temporal scales. It's a challenge that comes and I think a very positive, very fruit-

ful challenge that we are faced with in our humanities disciplines, as I said, to get out of our comfort zone. We already knew this is not the only temporal scale that we should deal with, but we hardly have the tools, the techniques, the instruments, analytic capabilities. So, the Anthropocene is really challenging us and what we are doing. I do see this as a positive challenge, but you can see easily and understand that there are others […] who look at this differently (Helmuth Trischler in Interview Zalasiewicz 2017).

The will to master this challenge exists. That's what the projects and approaches mentioned show. Initiatives such as IHOPE or Stewart Brand's *Clock of the Long Now* impressively demonstrate that such a challenge can be met. One way of approaching the task is to collaborate with geologists.

The project of building a clock, which is supposed to be able to run for 10,000 or more years independently of human intervention, initiated by the Californian utopian Steward Brand, impressively points to the dialectical constellation that humanity has to assume responsibility for the consequences of its actions that extend into an almost endlessly distant future, without knowing how it can meet this responsibility (Hillis et al. 2011; The Long Now Foundation 2019). Projects like the *Clock of the Long Now* give voice to the hope of leaving behind categorical boundaries that emerged with the Enlightenment. In this way, the merging of geological and historical time becomes feasible.

One of the interviewed historians describes the need for *Vergleichzeitigung* (simultanization) of the different time scales by stating: "[W]hat I think synchronizers do […] through the Anthropocene, they're going to link between the geological time scale, and the agency of humanity in the world, and then also linking the individual human life into the temporalities of the planetary, and the temporalities of the environment, and the temporalities that are ultimately geological. […] A new logic and also directionality comes into time" (Interview B8, 2018).[7] A recently published edited volume by a group of historians attests to new conceptual directions and potential new narratives based on the multitemporality of the Anthropocene (Ekström and Bergwik 2022; cf. as well Jordheim 2022).

Another way of approaching the task of conceptualizing new temporal narratives is to collaborate with geologists. Instead of narrating historical change on the horizontal level, it is important both to develop nonlinear narratives and to locate history on the vertical level (Trischler and Will 2019; Westermann and Höhler 2020). Stratigraphers read Earth history vertically, analyzing geological strata to identify disruptive phases of planetary scale alongside units of time of continuous change. In contrast, history is invited to further develop Ernst Bloch's idea of the simultaneity of the non-simultaneous in order to conceptualize layered models of historical temporality— "one thing is to split your mind and […] to keep part of your thinking in

[7] The concept of the synchronization of time goes back to the cultural historian Helge Jordheim, who describes two central instruments of historicization: First, progress as an instrument of modernity that created homogeneous, global time. According to Jordheim, progress is now being replaced by, secondly, crisis. While progress is oriented toward the future and thus characterized by continuity, crisis is oriented toward the present and synchronized via discontinuity and rupture; cf. Jordheim (2014, 2017), Jordheim and Wigen (2018), Sörlin (2022).

the time scale of individual humans and other nonhumans, and then deep time as well" (Interview Szerszynski 2018; see also Interview B2, 2018). Libby Robin draws a similar conclusion. Accordingly, particularly the aspect of deep time that becomes apparent in the Anthropocene requires a translation into other formats: "[W]e need to translate it into the imaginative, the aesthetic, the gut feeling, the affective, all those things are needed, and that's where the humanities and the arts and the museums come in actually" (Interview Robin 2017).

Environmental historian Sverker Sörlin describes the task of the historian in 2018 as follows:

> We need the history of nature in order to write history. We need the stories and the big drama and the Earth histories of the sciences. And we can't leave these stories to scientists alone, because their stories lack details of the human. [...] Now is the time to use our skills. We will expand in many directions and we will go to places we have never gone before (Sörlin 2018b, p. 23).

And this is exactly what highlights the complementary potential of a study of the Anthropocene beyond disciplinary boundaries. The seeming incompatibility due to different disciplinary methods and epistemological interests dissolves in favor of manifold anthropocene realities. The negotiation process within the AWG about the (in)compatibility of geological, archaeological, and historical temporality, which becomes explicit in the discussions about the birth date of the Anthropocene, illustrates this point (see Will 2021, pp. 144ff.). The classical historical conception of time differs from that of geology and archaeology not only in terms of its temporal capacity. Moreover, history follows a completely different logic beyond materially spatialized layers of time. Historically relevant things are often insignificant for geologists: "[C]ompare the scale of earth in history and that, then humans are a little passing blip" (Interview Zalasiewicz 2017). Besides that, the geological practice of decision-making on the one hand and the historical practice of interpretation on the other hand are fundamentally different from each other. While the geological process of voting suggests unambiguity, the process of interpretation and perspectivization in historical sciences opens up space for ambiguity and periodizations that coexist. Global historian John McNeill describes this tense coexistence between the two methodologies as follows:

> I am consistently amused by the radically different procedures that historians operate by and geologists operate by. [...] [G]eologists have very formal procedures and [compared to a historian] you cannot simply say as a geologist that we're now in the Anthropocene. [In history] [...] anybody can say anything about periodization [...]. We are content to let anarchy reign, and content to argue endlessly about proper intervals, and to recognize different intervals for different themes, different intervals for different parts of the world (Interview McNeill 2017).

Due to the lack of comparability, McNeill doesn't see any incompatibility between the two disciplines. The fundamentally different epistemological interest inevitably requires different methods (see also Interview Waters 2017). Even though the meanings of the concept have multiplied and sometimes strongly diverge, the common motivation has remained: to understand the Anthropocene and to gather evidence for the Earth systemic change caused by anthropogenic action. In the

Anthropocene, the different timescales converge in novel, synergistic ways—not least geological history itself, once described by Fernand Braudel as moving slowly and steadily across long timescales, becomes fast and experienceable through eventful changes in the Anthropocene, characteristics that Braudel once attributed to the historical time, the *histoire événementielle* (Braudel 1992). This encourages the disciplines involved to enter into an active exchange with each other and to enable new orientations by means of new, complementary models of temporality, which the Anthropocene virtually demands. The interview material as well as observations at the 2018 AWG meeting in Mainz show that the anthropocene geological-historical interference is welcome and partly even intended by geological actors (Damianos and Waters 2018; Interviews McNeill 2017; Williams 2017; A4*, 2017; A3*, 2017; A1*, 2017; Zalasiewicz 2017). By deliberately referring to the historical significance of the provisionally consented beginning in the 1950s, geological actors succeed in mobilizing knowledge from outside their discipline as an epistemic resource for their own argument as the following quote shows:

> However, a boundary placed at the time instant of the Alamogordo test would mark a historic turning point of global significance associated with the Great Acceleration, while in practical stratigraphic terms it would include all primary stratigraphic signals of bomb-related radionuclides, including those of the geologically simultaneous Hiroshima August 6, 1945 – and Nagasaki – August 9, 1945 – bombs (Zalasiewicz et al. 2015, p. 200f.; Head et al. 2023).

Examples like that illustrate that the bringing together of not only different time scales but also different disciplinary expectations is feasible. At the same time, however, it is exactly this coincidence between the chosen birth date of the Anthropocene with the primary marker plutonium 239 and its historical significance that raises concerns among some scholars. Emlyn Koster discusses the possible negative implications of the often called "bomb spike" "given the world's abhorrence of nuclear war" (Koster 2023).

4.3.2 *Placing the Planetary*

The consideration of temporal perceptions as well as of the relation between nature and culture from cultures of the Global South also holds great potential in this respect. Thus, some historians have looked at Asian cultures by asking for new historical narratives with regard to the possible consequences arising from the current state of planet Earth. According to Julia Adeney Thomas, what is cited as a reason for the omission of, for example, Asian processes in traditional historiography—namely, that freedom from the constraints of nature became a defining feature of modern historiography—is now being considered as a solution strategy for the required shift in the Western conception of nature-culture in the Anthropocene (Thomas 2017a, 2017b): the eco-spirituality practiced in Asian religions such as Shintoism or Buddhism has always been guided by the notion that human life is embedded in nature. This conception gives particular voice to Asian perspectives now: both in relation to discussions about adequate ways of dealing with anthropocene vulnerabilities and with regard to strategies for enhancing the capacity for resilience in face of global threats (Hudson 2014; Bruno 2022). The same applies to cultural and religious practices from Latin America. The cult of *pachamama* among indigenous people, for example, stands for a life in harmony with nature. It is Latin American countries that also take a pioneering role with regard to the legal anchoring of the rights of nature. In 2020, Costa Rica granted citizenship to bees and other pollinator species, in spring 2018, Colombia's Supreme Court caused a worldwide sensation as it declared the Colombian Amazon a legal subject (República de Colombia, Corte Suprema de Justicia 2018; Greenfield 2020). Ten years earlier already, the Ecuadorian constitution granted Pachamama the right to rehabilitation, among other rights (National Assembly, Legislative and Oversight Committee 2008; Etchart 2022). In 2017, indigeneous groups from both New Zealand and India succeeded in making two rivers, Whanganui River and Gangotri and Yamunotri Glaciers, subjects of right (Knauß 2018). In addition to the question of legally anchoring the rights of nonhuman entities, this puts the issue of intergenerational responsibility at stake. Environmental law, which offers space for both of these perspectives, has further evolved as a discipline as well as in practice in recent years (See e.g. Challe 2022; Darpö 2021). With Davor Vidas, who is specialized in law of the sea, this perspective is present also in the Anthropocene Working Group.

Thomas demonstrates pars pro toto that widening the perspective beyond specific geographical space can be productive. She leaves the comfort zone of her own discipline and enters an active negotiation process about the concept of the age of humans by using biological contents as a reference (Thomas 2014). Furthermore, in her studies on manifestations of the Anthropocene in Asia, especially in Japan, she takes up the demand for localizing the planetary, the "placing, situating the Anthropocene" (Interview Pádua, 2017), which has been expressed, among others, by the anthropologist Anna Tsing and the historian José Augusto Pádua (Austin 2017; Elverskog 2014; Ghosh 2017; Marquard 2019; Pádua 2017; Tsing 2015; Interview Pádua, 2017; Interview B1 2018; Tsing 2018).

The examination of the Anthropocene from a local perspective needs a transformation of scalar tools and conceptions, which historical sciences attempt to implement by linking the planetary with the global, the national, and the local through its localization. Historian Libby Robin puts it as follows: "I think that scaling up and scaling down is the biggest challenge, and it's true of every aspect of the Anthropocene, it's not just the time. [...] I'm really interested in [...] localizing the Anthropocene, because it's so big and so much you feel you can't do anything about it so you just cripple. But [...] the Anthropocene happens in local places too. And if you make it local, you suddenly make different people experts, and you enfranchise a whole lot of action" (Interview Robin 2017).

A look into the past shows that bringing together formerly separated scales is a feasible task and that the different dimensions may well be compatible with each other. Indeed, this is not the first time that people in both the scientific and non-scientific spheres have been faced with the challenge of abandoning established conceptions of space and time. Similarly, in the nineteenth century, when the biblical view of space and time was extended by the geological, spatially expansive deep time as a result of the further development of dating methods, scientists have been challenged to develop new narratives and models of historical temporality.[8] Admittedly, both the tools of traditional historical methodology and the established linear narratives may no longer be sufficient for such an Anthropocene historiography. The histories of placing that will be examined in the following illustrate that this approach by no means results in homogenization—a concern raised by numerous historians with regard to macrohistories as well as the use of big categories such as the planetary and global.

The local approach to the Anthropocene concept is a task that historians can master in particular, as they are able to create and establish such a connection on the basis of their methodological approach and their epistemological interest. The potential to connect diachronous, local Anthropocenes with the synchronous, planetary geological concept makes historians an active, indispensable part of the debate—complementary to the geoscientific negotiation of the concept and in a way completing the natural scientific evidence. Translating single aspects of the geological concept and making it comprehensible by the practice of placing has a kind of feedback effect on the differentiated perspective sought by the humanities and social sciences. Furthermore, this also offers an opportunity for legitimization for humanities disciplines, the historical sciences in particular. According to Chris Otter, addressing the Anthropocene concept in the historical sciences has "significantly widened and opened the scope of our field" (Otter et al. 2018, p. 596).

The localization of anthropocene themes finally offers historians, in contrast to geologists, the possibility of conveying political messages—messages that take up conceptual criticism of the Anthropocene and dispel it to a certain extent. The localization of anthropocene phenomena outside the Western world holds the possibility of opening up a differentiated picture that, in its complexity, in turn takes into account the planetary claim of the Anthropocene concept. Historical science is always also political practice—not least because the view into the past always opens up a view into the future as well (Will 2021).

Historians can draw from a wealth of material in this regard. Although the existing literature on placed Anthropocenes is still limited, different attempts of localization already exist. While some choose a continental focus, others opt for a national or regional perspective. Studies on an Asian or African Anthropocene as well as on a Chilean, Japanese, or British Anthropocene have, for example, been carried out so far (Otter et al. 2018; Issberner and Lena 2017; Pádua 2017). Historian José Augusto Pádua approaches the Anthropocene from a Brazilian perspective. He specifically examines Brazil's role in three phases of the Anthropocene: from industrialization to 1950, for the period of the Great Acceleration from 1950 to the present, and for a future period of a "Self-Conscious Anthropocene." In his view, such a history has the potential to tell an "encompassing, though synthetic history of the country [...]

[8] The revaluation of space already took place with the Enlightenment around 1700. A classic work on this subject, which has, however, also been much criticized, is Koyré (2008). With the Theory of Relativity the spatial scale expanded again around 1900. Cf. as well Coen (2016).

[that] might demonstrate the relevance of thinking national histories from a planetary perspective" (Pádua 2017, p. 24).

As a recent publication from the field of *urban studies* shows, the concept is beginning to be used in research on urban spaces as well (Thomson and Newman 2018; Ljungkvist et al. 2010). Regardless of the perspective we choose—localization presents itself as a productive option for addressing the challenge of scaling posed by the Anthropocene. The way Gabrielle Hecht approaches the topic offers an instructive example. Taking the uranium-bearing rock of an open-pit mine in Gabon as an empirical object that displays the characteristics of a natural reactor, she outlines a theory of the interscalar. Hecht regards the understanding of certain objects as interscalar vehicles as a way to handle the multidimensional spatial and temporal scales of the Anthropocene. Every *scale* is at the same time a category and as such the result of sociocultural and/or political processes (Coen 2016). Consequently, one of the characteristics of scales is also that they are always in the process of becoming, which encourages Hecht to argue for the use of objects as interscalar narrative tools in historical studies (Hecht 2018, 2023). She links this to the objectives of developing a novel anthropocene geohistorical spatiotemporal scale and of rendering geohistoricity more tangible through the materiality of the interscalar object. The broiler as a biofossil, plastic as a new geological material, and the new category of technofossils can be attributed a similar function (Will 2021). Plastic, for example, as an omnipresent object of everyday life, makes the geologicity of the Anthropocene literally tangible. It thus unites the historical and geological past, present, and future. Culture becomes a part of geological nature via the sedimentation of plastic in stratigraphic layers and a factor of human and animal nature via the ingestion of food that contains plastic particles. Donna Haraway's *nature-culture* proves to be an appropriate label in this context. The same applies to the broiler chicken and technofossils. All three of these object categories take a bridging role with regard to the various temporal scales (Will 2021). Timothy Morton's remarks on *hyperobjects* follow a similar path.

Hecht's interscalar approach offers great potential in dealing with the planetary dimension of the Anthropocene concept, and she convincingly illustrates that with the example of Oklo. The biologist Colin Hoag and the anthropologist Jens-Christian Svenning take a quite different approach in their examination of Africa's role in the Anthropocene. Committed to the aim of drawing conclusions for human adaptation possibilities from the knowledge they have gained, they explore how indigenous African communities dealt with environmental change over a period of about 2.6 million years. Like Mark Hudson for the Asian Anthropocene, they make a strong case for the local African past as a means of dealing with a planetary future (Hoag and Svenning 2017; Hudson 2014).

Geographer Kathryn Yusoff has a very different take. In her 2018 book *A Billion Black Anthropocenes or None*, she takes geographically based inequalities between Western and non-Western nations into focus, constructing "the proximity of black and brown bodies [...] [as] an inhuman proximity organized by historical geographies of extraction, grammars of geology, imperial global geographies, and contemporary environmental racism. It is predicated on the presumed absorbent qualities of black and brown bodies to take up the body burdens of exposure to toxicities and to buffer the violence of the earth. Literally stretching black and brown bodies across the seismic fault lines of the Earth, Black Anthropocenes subtend White Geology as a material stratum" (Yusoff 2018, p. vii).

The localization efforts only briefly outlined here show that the intended global differentiation is primarily interested in dissolving the eurocentric focus of the debate. The path of global spatialization taken by local appropriation demonstrates worldwide spatiotemporal interdependencies in a historical-geological perspective and thus underpins the planetary being of the Anthropocene (see, e.g., Waters et al. 2022). It testifies to a certain paradox that placemaking, which also represents differentiation, ultimately underpins the planetary dimension of the phenomenon and thus the geological thesis.

4.4 Ambiguity of Previous Unambiguities

The discussion among humanities scholars and social scientists about responsibilities, agency, and the question of who actually counts as a subject or object in the Anthropocene also testifies to the fact that established boundaries between categories are blurring in ways that have not been seen for a long time. Previously clear analytical frameworks and conceptual categories have become ambiguous.

In addition to the *anthropos* as according to the widespread and long-lasting conception of central acting subject, in the Anthropocene, other species and things step out with subject characteristics, which were formerly regarded as objects. These include plants and animals as well as materials or technology. In the following, the chapter will explore, specifically based on the example of technofossils, that this blurring of boundaries offers potential to make the interscalarity of the Anthropocene more tangible.

A second ambiguity, which, due to the effect-centeredness of stratigraphy, does not play a major role in the geological debate, emerges in the discourse about responsibilities in the Anthropocene. This discourse occurs on different levels of differentiation and is ultimately committed to the goal of adding causal relationships and developments that preceded over centuries to the geological Anthropocene. The discussions about responsibilities bring to the fore the methodological differences between disciplines in the natural sciences and the humanities, which manifests itself in relation to the negotiation of the Anthropocene primarily in the cause and process orientation of representatives of disciplines in the humanities.

The starting point of the discussion about the so-called Capitalocene, for example, is the criticism often formulated by humanities and social science scholars that the Anthropocene concept, by terminologically mirroring the species category in the determinant *anthropo*, obscures both inequalities and different dimensions of responsibility. Instead of stimulating urgently needed social and political changes, the concept would obscure the attribution of concrete responsibilities that are closely related to capitalist interests.

The discussion of capitalism already contains agency attributions to nonhuman entities, in this case capital, which show parallels to posthumanist approaches. Representatives of posthumanism feel particularly provoked by the Anthropocene concept. They fear, for example, that the negotiation of the Anthropocene further promotes the anthropocentric perspective they reject, instead of counteracting it, which they consider so important in view of the phenomena the Anthropocene describes. The posthumanist discourse here follows two approaches: While some focus on nonhuman living beings such as plants and animals, others begin to explore the agency potential of material entities.

Another challenge posed by With the Anthropocene, thus, the question of who or what counts as an agent in the Anthropocene besides the *anthropos* has come to the fore again. In this context, an expansion of the spectrum of actors is looming that is quite significant—especially for history as a science that has been methodologically focused on humans as the central actor until now. A similarity to the new perspective entering geology, away from materials and nonhuman actors toward the human species, emerges here. This impressively demonstrates that the two cultures ultimately struggle with similar difficulties, suggesting that experimental, collaborative settings have the potential to enable productive interactions in the future.

Even more so than in the natural sciences, there seems to be a growing awareness in the humanities of the need for a new kind of science, "which is not only the sum of the individual parts of different disciplines" (Interview B5, 2018, originally in German). What the Anthropocene needs instead is entangeled expertise. This does not imply a consensus on what the Anthropocene means. Instead, the Anthropocene can work as a multiple concept in which different perspectives complement each other and merge into an overall picture. This reveals a special potential of the Anthropocene concept, which Tim LeCain puts as follows: "[T]he key [...], the beauty of the Anthropocene in a more narrowly constrained way, is the bringing together of science and humanities" (Interview LeCain 2018).

Things that seemed unambiguous for a long time have become ambiguous—boundaries between established categories are becoming just as blurred as the epistemic boundaries between the disciplines involved. This results in an interaction between collaboration and competition, which promises to spur new ideas.

4.4.1 Technology as a Bridge Across Time and Space in the Anthropocene

An instructive example for the dissolution of the classical subject-object dualism is the discourse on the role of technology in the Anthropocene. Moreover, the discussion sheds light on possibilities of connecting the deep geological past and the human time in a synergetic way.

Research in both technology pasts and technology futures related to the periodization, the history, or future challenges of the Anthropocene has been part of almost every single aspect discussed under the concept of the Anthropocene so far—across the numerous disciplines involved in the debate. It is no coincidence that several of the proposals for an alternative naming of the Anthropocene situate technology center-stage. And it was the eminent force resulting from new technologies such as the steam engine that empowered humanity to transform the Earth and bring (among other factors) about the conditions of the Anthropocene (Will 2021; Trischler and Will 2017, 2019).

The debate about the role of technology in the Anthropocene brings together different conceptions of technology and scrutinizes established narratives. While some argue for critically questioning the promise of progress and the certainty of salvation through technology—not least because human responsibility for earth-systemic changes is thus shifted to technology (see e.g. Hornborg 2015; Soriano 2018; Will 2021, pp. 183ff.)—creeds in the technofix are experiencing a renaissance in the course of the debates on the use of geoengineering in dealing with climate change (Oomen 2021).

From the 1970s onward, there has been a consensus in both STS and history of technology that technology is shaped by society (see, for example, Bijker et al. 1987; Williams 1993). By introducing the concept of the technosphere, geoscientist Peter K. Haff, member of the Anthropocene Working Group, has fundamentally challenged this view. For him, technology does not result from human actions, but rather from a dynamic Earth system, the technosphere, which humans cannot control, which they cannot escape from, and in whose service they act. Therefore, human beings are "locked into technospheric feedback loops of performance and provision [...], from which position they cannot exit without peril to necessities of existence" (Haff 2017, p. 108). The technosphere, defined as "the interlinked set of communication, transportation, bureaucratic and other systems that act to metabolize fossil fuels and other energy resources," acts autonomously and according to its own system-specific rules (Haff 2014b, p. 301). It is to be located as a new emerging Earth system on the same level as the atmosphere, the lithosphere, the biosphere, and the hydrosphere (Haff 2014a). Haff thus builds on older debates about "autonomous" technology that have pointed to the complexity of a decoupled technical sphere, but he goes one big step further by conceptualizing it at the systems level. His non-anthropocentric perspective repositions humankind on the planet by focusing on the the Earth systems level.

Despite all the criticism Haff's concept has met (see, e.g., Donges et al. 2017; Edwards 2017; Lorimer 2017; Renn 2020, pp. 382ff.; Interview A4*, 2017), thinking in technospheric terms can *firstly* contribute to a better understanding of the temporal and spatial geologicity of technology and human activity. The temporal scale of technical development thereby not only expands in the direction of big history and deep history but also gains a physical presence through its material manifestations. The Anthropocene has forced stratigraphers to develop a new method and field, namely technostratigraphy, which extends the life cycles of current and future technologies into a distant future, when the physical remains of these technologies will have fossilized, thus creating new strata of technofossils. Those technofossils inherit the potential to function as a possible bridge between the vertically running geological deep-space-time and the horizontally running human time. The beginning of technofossil deposits, defined as "structures made by species of Homo" (Zalasiewicz et al. 2019b, p. 144), is set with the earliest biostratigraphic record of Homo about 2.8 million years ago. Since then, technofossil diversity has continued to grow. With the Great Acceleration, this development became exponential, and human-technology interaction reached a scale that blurred the boundaries between nature and technology, even at the fossil level. The bones of broiler chickens, which belong to the category of biofossils but whose structure and chemical composition result from the technical possibilities of breeding farms (which makes them appear more as a techno-biofossil hybrid than as a classical biofossil), are an example of this. Technofossils appear as interscalar objects in several respects, connecting the past with the present and the distant future as well as the horizontal history of humanity with the vertical history of the Earth. The concept of the technosphere thus supplements the previous spectrum of (techno)historical temporality by its deep-time geological-physical dimension. These new temporalities connect human temporal scales with geological timescales, deep pasts with deep futures. They challenge the existing theoretical concepts and methodologies of studying technologies futures and ask for new epistemologies tool to bridge multiple temporal scales.

This results, *secondly*, in a replacement of linear conceptions of time and challenges historians, especially historians of technology, to conceptualize layered models of historical temporality. Established narratives are put into question, which consequently confronts historians of technology with the challenge of rethinking the fit and explanatory power of those narratives. This is especially true for the conception of technology futures, which are to a large extent based on knowledge of past historical trajectories. With the observed extension of the temporal and spatial spectrum, however, futures are becoming even more open and ambiguous, thus strengthening the need for interdisciplinary collaborations.

In the Anthropocene, and especially the discussions concerning the concept of the technosphere, *thirdly*, the subject-object dualism dissolves in favor of a multiplication of itself: thus, humankind as a currently acting subject not only unites the geological, historical, and biological past, present, and future in itself, but in a technospheric perspective—as a subject acting in the service of technology—it is also to be understood as a technological subject. At the same time, Haff's idea of autonomy with regard to technology denies humans active agency and turns them into technological objects in a system in which technology takes on the role of the acting subject. Consequently, it applies the reverse of what history of technology, STS, and related fields have consensually advocated since the 1970s. While this is often perceived as a challenge and provocation (LeCain 2015), it also holds great analytical potential, since humans, in their conception as both technological subject and object, act as a bridge between different scales of space and time. Technology thus proves—and this is an epistemological advantage not to be underestimated—to be a material as well as temporal bridge accessible to the human mind (between geological dimensions and historical dimensions as well as between the planetary scale and the human scale), which explains the interdisciplinary scope of this discourse.

There are some humanities and social science scholars who take a similar approach. While the historian Gabrielle Hecht with her interscalar vehicles and the sociologist Timothy Morton with his boundary objects also focus on objects as instruments that transcend time and space, the philosopher Clive Hamilton places the emphasis in his reflections on humans as embedded subjects, relating to concepts of Gaia as well as the conception of the technosphere or the ergosphere. It is no coincidence that the categories of subject and object are undergoing a transformation. According to social ethicist Markus Vogt, these categories, along with the addressee, are precisely those dimensions of responsibility that are unsettled in the face of planetary anthropocene phenomena.

What can we learn from anthropocenic conceptions of technology and how applicable they are with regard to the conception of technological futures? The concept of the technosphere in its complexity points to the fact that we need to deviate from established paths in conceptualizing technology futures in the context of the Anthropocene—and this means, not least, finding a way to deal with wicked problems when it comes to ensuring the habitability of planet Earth for future generations: *firstly*, there is a need for the interplay of different approaches—to find a way of dealing with the challenges that lie ahead of us and, moreover, to make them communicable and understandable to different societal groups (to think in terms of sociology of expectations is, for instance, as important as reaching the political sphere by thinking in line with the concept of sociotechnical imaginaries). *Secondly*, spatial and temporal dimensions merge, which challenges historians of technology to develop new narratives based on the changing temporalities as well as the changing actor constellations.

The fact that technology futures also become more comprehensible through the described bridging effect on the one hand and their material presence on the other entails the potential to involve the public through science communication formats in new ways. Historian Chris Otter recently made the case for "[s]ocializing the technosphere" (Otter 2022, p. 953). There is hardly anything our society needs more than a multilateral dialog.

4.4.2 New Responsibilities

Who Is the Anthropos in the Anthropocene?

In the discussions about the *anthropos* as a conceptual category, tensions emerge that stem from the methodological differences between the natural sciences and the humanities and social sciences.

The open discourse on this topic began in 2013, when the sociologist and environmental historian Jason Moore introduced the term Capitalocene as a counter-proposal to the Anthropocene that addresses the heterogeneity of humankind.[9] The source of Moore's discussion of the Anthropocene in the context of capitalism research was the criticism often formulated by the humanities and social sciences that the Anthropocene concept obscures differences, inequalities, and different levels of responsibility by conceptualizing humanity as an undifferentiated, homogeneous entity that is the driving force behind the planetary crisis (cf., e.g., Baskin 2015; Manemann 2014; Hartley 2016; for a critical discussion of Moore's theses cf., e.g., Watson et al. 2016; Angus 2016; Nayeri 2016; Angus and Foster 2016; Albritton Jonsson and Wennerlind 2023).

According to their disciplinary logic, natural scientists operate with the concept of species. In geology, this is due to the scale of deep time. Geological transformations that occur over millions of years show a connection with the presence or absence of species, but not of specific varieties of certain subspecies. Many humanists and social scientists, however, see the use of the species category for the age of humans, like Moore, as an untenable reduction in complexity (the concept of species as such is also a controversial biological term, which is the subject of a separate debate; cf., e.g., Ereshefsky 2016). From a historical point of view, this is true. While the one, in accordance with disciplinary logic, resists unifying thinking, the other's methodology resists differentiating.

Nevertheless, the Anthropocene confronts the humanities and social sciences with the challenge not only to go beyond their previous investigation period but also to overcome corresponding ways of thinking. For it is only the deep time perspective that allows us to see why the current environmental changes, which result in Earth system ruptures, pose a long-term threat to humans and indeed to the human species. It runs particularly against the methodology of history to look at humanity as an undifferentiated entity, since this obscures causal details of historical change and is in tension with the complexity of the conception of human-nature relations.

Even the geographer and political scientist Kathleen McAfee, who does neither question the concept of species as such nor Chakrabarty's argument that humans only became a geological factor as a species, emphasizes that the decisions of power-bearing individuals were nevertheless the decisive criterion with regard to the geologization of humans—whereby this power of individuals was in turn the result of political and social processes (McAfee 2016). This part of the debate particularly focuses on the questions: *Who is the we?* and *Whose Anthropocene?*

While the causes with regard to those developments that led us into the Anthropocene are to be found in a differentiated analysis of single, interacting microlevels, their effects show up in a planetary-collective manner on the macrolevel. Since representatives of geology repeatedly state that it is the task of the humanities and social sciences to uncover the causal relationships that gave rise to the effects examined by the natural sciences, the species category is in fact hardly

[9] According to Donna Haraway, Andreas Malm first came up with the term Capitalocene in a seminar at Lund back in 2009. Moore also points out that he adopted the term from Malm; cf. Haraway (2015); Moore (2018); David Ruccio also uses the term Capitalocene even before Jason Moore in a 2011 blog post Ruccio (n.d.).

suitable for a critical discussion of the Anthropocene concept in the humanities (Head 2014; Sideris 2016). Nevertheless, an open attitude toward species thinking would promise *one* opportunity to establish a connection toward the geological deep time and thereby overcome the separation between nature and culture. Moreover, de-differentiation in this particular case presents itself as a way to think in planetary terms that transcends the historical conception of time, even from a historiographical perspective. If this happens reflexively, the species perspective can help history to conceptualize new models of historical temporality in a planetary perspective, which—once established—can be filled in a differentiating way.

By including this comprehensive category in the historical sciences, Chakrabarty sees potential. Among other things, the chance to supplement established modes of thinking and to contribute to a better understanding of one's own geologicity, for "[w]e may not experience ourselves as a geological agent, but we appear to have become one at the level of the species. And without that knowledge that defies historical understanding there is no making sense of the current crisis that affects us all" (Chakrabarty 2009b, p. 221). Geographer Gerda Roelvink too explores the potential of integrating the species category. Her work points in a similar direction (Roelvink 2013).

Understanding the crisis goes hand in hand with recognizing a (environmental) responsibility that is not only intergenerational, but rather delimited and planetary. The concept of responsibility, as we understand it today, is a term that has its roots in fifteenth century legal texts. There are three essential elements that constitute the term: responsibility is a competence that lies (1) with someone, (2) for something, (3) toward someone (Vogt 2019, p. 39). If we take seriously the differentiation of mankind in relation to the causes of the Anthropocene, which has been called for frequently by the humanities, it would be wrong to assign the same degree of responsibility to all people on the way to safeguarding a functioning Earth system (on the question of Responsibility in the Anthropocene cf., e.g., Kistler 2018). Political scientist Jeremy Baskin revisits this criticism in a recently published book chapter. He criticizes with regard to the debate about the Anthropocene as a geological concept the "repeated listing of symptoms of anthropogenic influence. There is little mention, however, of what might be driving the Anthropocene. [...] Justice questions and inequality go unexplored" (Baskin 2019, pp. 154ff.). Instead of repeatedly criticizing geology for this supposed failure, it would probably be more useful to develop solution strategies within their own ranks for a responsible way of adressing the challenges of the age of humans. Indeed, according to Markus Vogt, in the face of the planetary phenomena described by the Anthropocene, all three dimensions of responsibility are unsettled: subject, object, and addressee. This manifests itself in an overload of responsibility and inhibits action. In supplementing the existing academic system with a transformative science, Reinhold Leinfelder, Markus Vogt, and many others see the potential for academia to become a catalyst for processes of social change by dissolving strict demarcations between the intra- and extra-scientific spheres (the concept of transformative science has sparked controversial discussions within academia about the merits of such inner-academic change; cf., e.g., Strohschneider 2014; Schneidewind 2015; Grunwald 2015; Renn 2020).

Agency

The concept of anthropocentrism describes the widespread belief in Western philosophy that humanity is a central entity superior to other forms of being. As a living being conceived as separate from nature, anthropocentrism ascribes a special moral position to humans. In this way, it justifies the exploitation of existing resources in the service of human well-being. While some consider the Anthropocene the first truly anti-anthropocentric concept (Morton 2014, p. 262), others believe "anthropocentrism puts the blame right where it should be" (Interview B1, 2018). And the category of the *anthropos* is criticized not only for its unifying character. Posthumanist thinkers, for example, fear that the negotiation of the Anthropocene concept further promotes the socially deeply rooted anthropocentric perspective instead of counteracting it; it would thus undermine the posthumanist way of thinking that has become increasingly popular from the 1990s onward. Cultural theorist Claire Colebrook puts it as follows: "From a modernity in which apparent difference was vanquished – acknowledging only life or matter without intrinsic difference – posthuman claims for being one more aspect of a general 'life' have now been vanquished" (Colebrook 2017, p. 6).

Especially, the natural scientific approach to human *agency* contributes to the reinforcement of this concern. Simon Lewis and Mark Maslin or Paul Crutzen, Will Steffen and John McNeill, for example, name the responsibility of humans in dealing with nature as the top priority when it comes to the management of planetary crises by exploring the question of humans as "Stewards of the Earth System?" (Steffen et al. 2007, p. 618). They note that "[t]o a large extent the future of the only place where life is known to exist is being determined by the actions of humans. Yet, the power that humans wield is unlike any other force of nature, because it is reflexive and therefore can be used, withdrawn or modified" (Lewis and Maslin 2015b, p. 178).

The decentering of humans aimed at in an abandonment of the nature-culture divide thus dissolves in favor of a multiplication of their centrality (cf. e.g. Baskin 2015). Humanity is not only the driving, causing force of the planetary crisis. In addition, in the person of a *planetary steward*, humanity is stylized as the savior who persists in nature-culture dualism (see e.g. Steffen et al. 2011; Steffen 2021). In reaction to the return of the traditional question about the (de)centrality of humans, a separate field of discussion between representatives of neo- and posthumanism has emerged in the debate about the Anthropocene as a cultural concept.

This subdiscourse is closely interrelated to other subdiscourses of the debate—just think of Haff's conception of an autonomous technosphere (cf. Trischler and Will 2017).

Since its beginnings, posthumanism has sought to overcome the dichotomous understanding of nature on the one hand and culture on the other and to understand humans not as separate from but as part of the ecosystem. If the Anthropocene concept pinpoints anything—it is the close interaction mechanisms between natural and social processes. At the same time, however, it provides "the most striking proof that humans do occupy a position separate from nature," which once again unfolds its provocative potential because it thus calls into question the very basis of the posthumanist approach (Hamilton 2017, p. 89).

> Contrary to the post-humanist stance, accepting the science of the Anthropocene entails accepting the unique and extraordinary power of humans to influence, by conscious decision or otherwise, the future course of the Earth, even if it is not necessarily in the way intended (Hamilton 2017, p. 89).

In response to the aforementioned characteristic of the Anthropocene concept, posthumanists take a counter stance. This manifests itself in an attempt to challenge the degree of human agency by shifting the focus to the agency of nonhuman entities. Accordingly, the planetary environmental crisis described by the Anthropocene should not be read as confirming the exceptionality of human agency, but rather demonstrates the ultimate failure of the historically grown hope of human control over nature and proves that the dualistic division into the categories of nature and culture has never existed outside our imagination. However, instead of leaving behind the underlying humanist values attributed to anthropocentrism, such as those of equality and dignity for all humans, posthumanists apply them to other species and materials. The hallmark of the redesigned ontologies is the decentering and, in a sense, deprivation of human agency in favor of an equal distribution of agency. It is a kind of paradox that this happens in the midst of the negotiation process about a concept that confirms the planetary agency of humans on the basis of scientific evidence as clear as never before.

Philosopher Jane Bennett, for example, who is primarily interested in "greener forms of human cultures and more attentive encounters between people-materialities and thing-materialities" (Bennett 2010, p. x), deprives humans of any intentionality when writing that

> "portions congeal into bodies, but not in a way that makes any one type the privileged site of agency. The source of effects is, rather, always an ontologically diverse assemblage of energies and bodies, of simple and complex bodies, of the physical and the physiological" (Bennett 2010, p. 117). "I believe in one matter-energy, the maker of things seen and unseen. I believe that this pluriverse is traversed by heterogeneities that are continually doing things" (Bennett 2010, originally in italics.).

Historian Tim LeCain, who develops his reflections on neomaterialism in *The Matter of History* from an excursus into microbiology, goes a step further in not only denying intentionality to humans, but in attributing not only agency, but also intentionality to things (LeCain 2015). From a neo-materialist perspective, culture and thus human beings are products of their material environment: "[H]umans are in the earth's hands" (LeCain 2015, p. 5). He proposes the term Carbocene in order to highlight the material history of the planet, especially the importance of coal and hydrocarbons.[10] With this proposal, he borrows from the Carboniferous, a geochronological period of the Paleozoic that began 359 million years BP (LeCain 2017, pp. 306ff.). By mobilizing geological knowledge through the analogy of the geological past of planet Earth, he succeeds in explicating the geologicity of humans. He combines scientific knowledge and historical reconstruction and comes to the conclusion that "the 'Anthropos'" are actually "the 'Carbos'" (LeCain 2017, p. 328) (on the interdependence between people and things cf. also Hodder 2012, pp. 15ff.).

Considering the stated intentionality of the material in this perspective it seems reasonable that the thesis of the external control of human activity by nonhuman agency draws a lot of critical reactions (Hamilton 2017, pp. 87ff.). In this perspective, humanity is relieved of any responsibility and freed from any guilt (see also the discussions about the Marxist fetish concept in the Anthropocene Hornborg 2013; Saito 2017; Cunha 2015; Malm and Hornborg 2014). However, the attribution of agency to the material simultaneously evokes an idea of shared agency and thus facilitates access to the geologicity of the human being: Locating humans at the geological level dissolves the categorical separation of nature and culture.

The feminist philosopher of science Donna Haraway, a particularly prominent representative of posthumanism, also takes the Anthropocene debate as an opportunity to further evolve her thoughts. Back in 1985 and 2003, Haraway published two manifestos, *A Cyborg Manifesto* and the *Companion Species Manifesto*, respectively. Both testify to her primary interest in the dissolution of established categories (Haraway 2010, pp. 149ff., 2003). Whereas in 1985, in accordance with posthumanist tradition, her focus was still on human-technology interaction, in 2003, she aimed at the perception of the inseparable interconnectedness of global and local naturecultures as well as the recognition of the agency of the *companion species*, comprising not only animals but also microorganisms and plants. In her critical discussion of specific implications of the Anthropocene concept, she not only takes a counter-position to the much-criticized anthropocentrism. Rather, she is concerned with the interplay at the systems level. Similar to other posthumanists, she leaves behind criticisms raised by the concept, such as the lack of contextualization and differentiation of humanity, and thus the attempt to construct causality in an explicative way. Instead, she operates at the actor level with diverse entities, material substances as well as living organisms, which on a conceptual level reveals an analogy to the discussion of the *anthropos* as a species—with the difference that she dissolves the centrality of the human species in favor of equally distributed agency.

The sociologist of science and philosopher Bruno Latour, who in his actor-network theory, consistently questioned the existence of established categories and has long left behind the idea of humans as the central actor, advanced his reflections, as does Haraway, against the background of the Anthropocene (Latour 2009, 2014b, c, 2015, 2016). According to Latour,

[10] LeCain is referring to coal and hydrocarbons.

> [t]he point of living in the epoch of the Anthropocene [...] that all agents share the same shape-changing destiny, a destiny that cannot be followed, documented, told, and represented by using any of the older traits associated with subjectivity or objectivity. Far from trying to ›reconcile‹ or ›combine‹ nature and society, the task, the crucial political task, is on the contrary to distribute agency as far and in as differentiated a way as possible – until, that is, we have thoroughly lost any relation between those two concepts of object and subject that are no longer of any interest any more except in a patrimonial sense (Latour 2014a, p. 15).

This is the only way to make the transition to an integrative, harmonious *geostory* (cf. here Latour 2014a).[11]

Since Haraway's work also aims to transcend categorical boundaries, it is not surprising that she dissolves small-scale categories in order to take a step further toward the blurring of boundaries at the big-scale level. Her much quoted plea, "make kin, not babies!" calls us to perceive ourselves in a relationship that transcends established boundaries between different entities and spacial and temporal boundaries (Haraway 2015, p. 162, 2016, p. 60).

> Cyborgs and companion species each bring together the human and non-human, the organic and technological, carbon and silicon, freedom and structure, history and myth, the rich and the poor, the state and the subject, diversity and depletion, modernity and postmodernity, and nature and culture in unexpected ways (Haraway 2003, p. 4).

With regard to posthumanist studies of the Anthropocene, Clive Hamilton notes a conflation of anthropocentrism as scientific fact, on the one hand, and anthropocentrism as normative assertion, on the other (Hamilton 2017, pp. 39ff.). Starting from the anthropocentrism as a scientific fact as declared in the debate on the Anthropocene as a geological concept—namely, that human activity has become such a strong impact factor within Earth system cycles, but without dominating or even controlling them, that has led the Earth system into an altered geological state—he constructs a new anthropocentrism as a kind of negative of teleological anthropocentrism. By means of four features, he exemplifies the novelty of his conception of anthropocentrism. It is based, first, on the recognition of humankind as an "embedded subject." In this way, he denies the exceptionality of mankind and its dominance over nature (Hamilton 2017, p. 52). While the new anthropocentrism, secondly, also assumes a human power capable of changing the planet, it draws ethical conclusions opposite to anthropocentrism. For, thirdly, while the previous discussion of anthropocentrism conceives of the Earth as a passive entity, the new anthropocentrism recognizes the Earth's activity and agency. In doing so, Hamilton, fourthly, in a way takes an anti-humanist standpoint—since he conceptualizes the

[11] Bruno Latour uses the term *geostory* in reference to and at the same time in distinction to Dipesh Chakrabarty, who operates with the term *geohistory*. Whereas Chakrabarty's term reflects the convergence of human history and geological history, Latour uses the term geostory to highlight the earth-bound nature of subject and object. Consequently, he is less concerned with the idea of reconnecting humans to nature than with depicting the connection of any kind of subject and object to planetary materiality.

fate of humanity as being dependent on many factors.[12] Instead of exploring normative implications related to anthropocentric thinking, he attempts to gain a constructive approach to the overall phenomenon of the Anthropocene by incorporating Earth systems science knowledge and lines of thinking. The inclusion of new categories, such as humans as *embedded subjects*, enables us to think of human agency across the different space-time dimensions in terms of its planetary nature (Hamilton 2016, 2017, pp. 36ff.). By referring to the recognition of human embeddedness in nature, he draws some positive lessons from posthumanism. At the same time, however, he emphasizes the need to remain aware of the singularity of human beings, which posthumanist approaches attempt to dissolve, primarily in terms of the possibility of influencing the future development of the Earth—not in the spirit of ecomodernism, but positively. He thus takes a position against the potential of posthumanist thinking in dealing with challenges of the anthropocene present and possible anthropocene futures, reversing the argument in the sense of recentering it.

A similar line of argumentation emerged from a philosophical-environmental science cooperation: In order to abolish the subordination of the Earth to human domination of traditional anthropocentrism, geographer Jeremy Schmidt and environmental scientists Peter Brown and Christopher Orr call for a reorientation in the direction of ethnocentrism. In the latter, different cultural perspectives and thus understandings of nature converge, linking Earth and human history. They thus argue for the creation of new concepts that meet the scale of planetary deep time without neglecting the historically differentiated perspective (Schmidt et al. 2016).

Instead of drawing a line under the debate about the Anthropocene as a geological concept, Hamilton's as well as Schmidt's, Brown's, and Orr's approaches toward the ethical consequences of anthropocene facts prove a great openness not only toward knowledge from different discplines but also toward the methodology that underlies this knowledge. Rather than selectively drawing on this knowledge to achieve a stabilizing and legitimizing effect for one's own perspective, Hamilton attempts to develop his new anthropocentrism starting from the Earth system science conception of the Anthropocene and to synergistically combine facts from natural sciences with the resulting normative implications. The same is true in reverse for Schmidt, Brown, and Orr.

4.5 Conclusion: The Anthropocene as an Academic Game Changer?

In the last two decades, the Anthropocene has developed into a concept with multiple meanings. The chapter has shown that the different understandings are not necessarily mutually exclusive. The Anthropocene debate seeks neither an exclusive approach nor homogeneous knowledge or one-dimensional perspectives. Rather, it

[12] I use the term antihumanism exclusively in reference to Hamilton's understanding of it.

is the complexity and the wealth of detail of the concept that bring its provocative potential to the fore—a provocation that, in addition to manifold discipline-specific manifestations, consists primarily in blurring established content-related, disciplinary, and categorical boundaries in a way unprecedented since the differentiation of disciplines in the nineteenth century. At first glance, the Anthropocene debate may appear to be a confusing mixture of divergent positions. A closer look, however, reveals that the overall debate consists of various subdiscourses. The lines of argumentation negotiated in these subdiscourses differ from one another in terms of their epistemological interest and consequently place different aspects at the center of interest. It is this diversity of relevant aspects that ultimately accounts for the claim of the Anthropocene. Thus, considered from a metaperspective, the subdiscourses do not stand in an oppositional relationship but in a complementary relation to each other and, in their sum, yield the problematic field described by the Anthropocene. An analysis of the debate not only sheds light on interaction mechanisms, commonalities, and differences between natural science and humanities and social science models of thought, but also on the limits of established processes of knowledge production (Will 2021).

Closing gaps of expertise within the ranks of the geosciences, contextualizing, differentiating, and perspectivizing the concept according to its ethical charge, as well as sounding out political, social, and economic implications, therein not only lies huge potential, but also a possibility for the legitimization of the humanities and social sciences in the Anthropocene. Mutual openness and rejection go hand in hand in this context. It is striking, however, that a large part of the central actors in the debate often not only faces the blurring of disciplinary boundaries with great openness but even intends this opening. Less centrally involved scientists, by contrast, react much more skeptically.

A conceptual novelty to which the chapter has paid particular attention is the intertwining of different temporal and spatial scales. However, we do not need to consider the very distant future of the Earth, nor the Big Bang almost 14 billion years ago as the beginning of everything, as propagated by representatives of "Big History," in order to recognize the perspective to link different timescales opened up by the Anthropocene debate (programmatic here Christian 2011a, b; Dukes 2013).[13] The Anthropocene concept allows us to link geological time and historical time into new narratives based on new temporalities (Trischler 2016; Robin 2015). Earth history and human history no longer run independently of each other, but coalesce into an integrated geo-history. Humanity shapes the Earth and is in turn shaped by geological signatures on which, in turn, humans have put their hands. This twofold interaction, accompanied by the blurring of the boundaries between nature and culture, has become the sign of the Anthropocene, which links different timescales. It links the long period of human intervention in the Earth that reaches back to the Neolithic Revolution with the present of the "long now" and the resulting responsibility for the future. Pointing out that geology emerged from the humanities, anthro-

[13] Cf. also the Big History Project supported by Bill Gates OER Project.

pologist Bronislaw Szerszynski formulates a paradox. The strict separation and seeming incompatibility of historical and geological time, he argues, are just another mental construct that also goes along with the categorical separation of nature and culture in modern thinking (Szerszynski 2017a, p. 116). This does not diminish the challenge and power of the Anthropocene to bring these two timescales back together. However, the feasibility of this is getting closer.

The chapter has shown that the entanglement of different timescales sets in motion the discipline-specific conceptions of temporality.

In this context, historians in particular face the task of integrating the planetary scale of space and time into their historiographical engagement with anthropocene topics. The tools to accomplish this are at their disposal. Narrativization, for example, through the practices of interweaving, translation, contextualization, (de)centering, differentiation and localization, enables them to take on the anthropological questions echoed in geoscientific discourse from within their own field of expertise. In this way, the humanities and social sciences have the opportunity to take on the urgently needed bridging function between the planetary and national and/or local scales of negotiated issues in order to make sense of the various aspects of the Anthropocene concept. So far, corresponding narrative models are lacking. Thus, historians are encouraged to conceptualize layered models of historical temporality that equally encompass deep geological time and related scientific phenomena as well as historical events. Drawing on several examples, the chapter has shown that this can be accomplished in collaborative settings. In particular, the history of science and technology, as well as environmental history have the opportunity to take on the role of an interdisciplinary and trans-disciplinary hinge in this dialog, to cast a critical light on established narratives, and to tell new stories on the basis of the new questions and new temporalities that result from the collision of historical and geological timescales. After all, interaction with sister disciplines of the natural sciences is part of the everyday business of these three subdisciplines of the history.

The same is true for other humanities and social sciences disciplines that deal with anthropocene questions. Be it literary sciences, which try to tell new stories based on new temporalities and altered understandings of the relationship between humans and nature, be it social ethicists, who are challenged to reassess questions of responsibility in the face of planetary challenges, or be it law where questions of nature as being a subject of right become more and more pressing in light of the Anthropocene, to name but a few.

To conclude, five areas of humanities response to the provocation of the Anthropocene and related challenges are accentuated:

Firstly, the different humanities disciplines produce innovative results in fields in which they have to closely cooperate with other sister disciplines. Yet, in order to fully utilize the epistemological potential of their field, the Anthropocene debate calls for a new conceptual and methodological step further and for seeking cooperation with the geosciences, in the spirit of the much-vaunted great interdisciplinarity (Costanza et al. 2007a; Robin and Steffen 2007).

Secondly, the connection between human and geological histories confronts humanities disciplines, especially historians, with the challenge of decoupling from outdated linear timescales to a greater extent than ever before. Instead of narrating historical change along a horizontal plane, historians have both to develop nonlinear narratives—which many have already convincingly demonstrated—and to locate history on a vertical axis. While stratigraphers read the Earth's history vertically, by looking at geological strata in order to identify phases of planetary-scale disruption as well as periods of normal change, historians have to conceptualize stratigraphic models of historical temporality (Armitage and Guldi 2014). The thoughts of historian Reinhart Koselleck thus seem more topical than ever. Koselleck already addressed the seeming incompatibility of scientific and historical timescales in his work *Zeitschichten* (Time Layers), published in 2000, and thus provoked a fundamental discussion among historians about the nature of history. In line with geology, whose concept of time is related to strata as material layers of time, Koselleck discusses the multilayered nature of historical time. It is no coincidence that he illustrates his reflections with examples situated around 1800, hence in the *Sattelzeit*—the period of time when the dichotomy of geological time on the one hand and historical time on the other hand, which characterizes modern thinking, emerged. Koselleck's plea for the inseparability of space and time as well as the inherent complexity of historical temporality now regains weight in view of anthropocene questions (Koselleck and Gadamer 2013; on conceptions of time in history, cf. also Graf 2012; Jordheim 2022).

Thirdly, the Anthropocene debate extends history as well as (more-than-human) agency not only in temporal dimensions but also in spatial ones. Global history has indicated ways of overcoming narrow nationalism and Eurocentrism to analyze transnational entanglements that focus in particular on the circulation of technological artiefacts and scientific knowledge. While global history operates at the *Flughöhe der Adler* (the altitude of an eagle), the Anthropocene concept charges us with the task of going further to take a planetary perspective on our history (Osterhammel 2017; Szerszynski 2017b).

It is one of the basic methodological and theoretical assumptions that every historiography is inevitably framed by the present day; and thus, current historical research is also Anthropocene history. Fourthly, however, Anthropocene narratives, be it ethical, historical, or literary narratives, are not only shaped by present-day concerns, but also by concerns about our future. They connect the long history of our present with implicit and explicit normative assumptions about our future, for example, with respect to the role of technology in solving problems that we have identified in our present and projected into the future. Exposing these normative dimensions of the Anthropocene debate in a multidimensional and thus trans-disciplinary way opens up perspectives that are both exciting and highly promising.

Fifthly, the provocation of the Anthropocene also manifests itself as a struggle for interpretive power. Stratigraphers complain vividly that the humanities and social sciences are depriving themselves of the *arcanum* of their field of research: the periodization of the history of the Earth. Their discipline risks to lose its epistemic strength and even to become a playground for political interests (Gibbard and

Walker 2014). Conversely, the humanities are also taking a stance: human history is the subject of cultural studies and whatever stratigraphers ultimately decide regarding the formal recognition and periodization of the Anthropocene is of little relevance to the humanities and certainly not binding. On the contrary, one of the primary tasks of humanities in approaching the Anthropocene is to develop alternative interpretations of the role of the *anthropos* as a subject of history, to unfold the processes that led into the Anthropocene, the different levels of responsibilities that come along with them, and the different local anthropocene realities and thus to fill the concept with more practical meaning through its localization (Hamilton et al. 2015).

Becoming mired in sterile professional political debate, however, takes a great deal of energy and ultimately helps no one. Instead, the potential of interdisciplinary cooperation and transdisciplinary dialog should be emphasized.

Only time will tell how things develop. On the basis of this chapter, however, it can be argued that the debate about the Anthropocene has set the course for the development of new lines of thought and new narratives, if not even of a new interdiscipline, which overcomes the classical dualisms of modern thought and operates beyond disciplinary boundaries (in more detail cf. Will 2021).

4.6 Bilbliography

4.6.1 Interviews

Coding of the anonymized interviews

A Natural scientists
B Humanities scholars and social scientists
C People at the interstice of science and the public
* Members of the Anthropocene Working Group

A1* 08.05.2017
A2 12.11.2018
A3* 15.11.2017
A4* 02.12.2017
B1 27.04.2018
B2 20.03.2018
B3 27.03.2018
B4 12.03.2018
B5 10.05.2018
B6 25.02.2018
B7 28.07.2017
B8 26.03.2018
C1 04.10.2017
C2 07.12.2017

Interview with Alf Hornborg. Lund/München 16.03.2018.
Interview with Tim LeCain. Bozeman/München 21.03.2018.
Interview with John R. McNeill. Zagreb 30.06.2017.
Interview with José Augusto Pádua. Zagreb 30.06.2017.
Interview with Libby Robin. München 28.11.2017.
Interview with Lise Sedrez. Zagreb 30.06.2017.
Interview with Will Steffen. Canberra/München 07.02.2018.
Interview with Bronislaw Szerszynski. Lancaster/München 28.02.2018.
Interview with Colin Waters. Leicester 20.06.2017.
Interview with Mark W. Williams. Leicester 20.06.2017.
Interview with Jan A. Zalasiewicz. Leicester 20.06.2017.
Interview with A1*. 08.05.2017.
Interview with A2.12.11.2018.
Interview with A3*. 15.11.2017.
Interview with A4*. 02.12.2017.
Interview with B1.27.04.2018.
Interview with B2.20.03.2018.
Interview with B3.27.03.2018.
Interview with B4.12.03.2018.
Interview with B5.10.05.2018.
Interview with B6.25.02.2018.
Interview with B7.28.07.2017.
Interview with B8.26.03.2018.
Interview with C1.04.10.2017.
Interview with C2.07.12.2017.

References

Albritton Jonsson, Fredrik, and Carl Wennerlind. 2023. *Scarcity. A history from the origins of capitalism to the climate crisis*. Cambridge, MA: Harvard University Press.

Angus, Ian. 2016. Knocking down straw figures. *International Socialist Review* 103.

Angus, Ian und John Bellamy Foster. 2016. In defense of ecological Marxism: John Bellamy Foster responds to a critic. https://climateandcapitalism.com/2016/06/06/in-defense-of-ecological-marxism-john-bellamy-foster-responds-to-a-critic/. Accessed 17 Aug 2020.

Antweiler, Christoph. 2022. *Anthropologie im Anthropozän. Theoriebausteine für das 21. Jahrhundert*. Darmstadt: wbg Academic.

Armitage, David, and Jo Guldi. 2014. *The history manifesto*. London: Cambridge University Press.

Asafu-Adjaye, John, Linus Blomqvist, Stuart Brand, Barry W. Brook, Ruth Defries, Erle C. Ellis, Christopher Foreman, David Keith, Martin W. Lewis, Mark Lynas, Ted Nordhaus, Roger Pielke, JR, Rachel Pritzker, Joyashree Roy, Mark Sagoff, Michael Shellenberger, Robert Stone und Peter Teague. 2015. Ecomodernist manifesto. Accessed 13 Sept 2017.

Austin, Gareth. 2017. Introduction. In *Economic development and environmental history in the Anthropocene: Perspectives on Asia and Africa*, ed. Gareth Austin, 1–22. London: Bloomsbury Academic.

Autin, Whitney J., and John M. Holbrook. 2012. Is the Anthropocene an issue of stratigraphy or pop culture? *GSA Today* 22 (7): 60–61. https://doi.org/10.1130/G153GW.1.

Baskin, Jeremy. 2015. Paradigm dressed as epoch: The ideology of the Anthropocene. *Environmental Values* 24 (1): 9–29.

———. 2019. Global justice and the Anthropocene: Reproducing a development story. In *Anthropocene encounters: New directions in Green political thinking*, ed. Frank Biermann and Eva Lövbrand, 150–168. Cambridge: Cambridge University Press.

Bauer, Andrew M., and Erle C. Ellis. 2018. The Anthropocene divide: Obscuring understanding of social-environmental change. *Current Anthropology* 59 (2): 209–227. https://doi.org/10.1086/697198.

Bennett, Carys E., Richard Thomas, Mark W. Williams, Jan A. Zalasiewicz, Matthew Edgeworth, Holly Miller, Ben Coles, Alison Foster, Emily J. Burton, and Upenyu Marume. 2018. The broiler chicken as a signal of a human reconfigured biosphere. *Royal Society Open Science* 5 (12): 1–11. https://doi.org/10.1098/rsos.180325.

Bennett, Jane. 2010. *Vibrant matter. A political ecology of things*. Durham: Duke University Press.

Bijker, Wiebe E., Thomas Parke Hughes, and Trevor Pinch, eds. 1987. *The social construction of technological systems. New directions in the sociology and history of technology*. Cambridge, MA, London: MIT Press.

Bösl, Elsbeth. 2017a. *Doing Ancient DNA. Zur Wissenschaftsgeschichte der aDNA-Forschung*. Bielefeld: Transcript Verlag.

———. 2017b. Zur Wissenschaftsgeschichte der aDNA-Forschung. *NTM* 25 (1): 99–142. https://doi.org/10.1007/s00048-017-0168-5.

Brauch, Hans Günter. 2021. The anthropocene concept in the natural and social sciences, the humanities and law—a bibliometric analysis and a qualitative interpretation (2000–2020). In *Paul J. Crutzen and the Anthropocene: A New Epoch in earth's history*, ed. Susanne Benner, Gregor Lax, Paul J. Crutzen, Ulrich Pöschl, Jos Lelieveld, and Hans Günter Brauch, 289–438., 1. Aufl. Springer eBook Collection, Bd. 1. Cham: Springer International Publishing; Imprint Springer.

Braudel, Fernand. 1992. Das Mittelmeer und die mediterrane Welt in der Epoche Philipps II, 2. Aufl. Frankfurt am Main: Suhrkamp.

Bruno, Andy. 2022. An Anthropocene history of Central Asia. *Kritika: Explorations in Russian and Eurasian History* 23 (2): 339–344.

Cahn, Michael. 1991. Wissenschaft und Literatur: Eine Berührungsstelle der zwei Kulturen. In Glanz und Elend der zwei Kulturen: Über die Verträglichkeit der Natur- und Geisteswissenschaften, eds. Helmut Bachmaier and Ernst Peter Fischer, 181–193. Konstanzer Bibliothek, Bd. 16. Konstanz: Univ.-Verl. Konstanz.

Castree, Noel. 2014. The Anthropocene and the environmental humanities. *Environmental Humanities* 5:233–260.

———. 2017. Speaking for the 'people disciplines': Global change science and its human dimensions. *The Anthropocene Review* 4 (3): 160–182. https://doi.org/10.1177/2053019617734249.

Chakrabarty, Dipesh. 2009a. *Provincializing Europe. Postcolonial thought and historical difference*. Princeton: Princeton University Press.

———. 2009b. The climate of history: Four theses. *Critical Inquiry* 35:197–222.

———. 2015. The Anthropocene and the convergence of histories. In *The Anthropocene and the global environmental crisis: Rethinking modernity in a new epoch*, ed. Christophe Bonneuil, François Gemenne, and Clive Hamilton, 44–56. London: Routledge.

———. 2016. Humanities in the Anthropocene: The crisis of an enduring Kantian fable. *New Literary History* 47 (2–3): 377–397. https://doi.org/10.1353/nlh.2016.0019.

———. 2021. *The climate of history in a planetary age*. Chicago, London: University of Chicago Press.

Challe, Tiffany. 2022. The rights of nature—can an ecosystem bear legal rights?. https://news.climate.columbia.edu/2021/04/22/rights-of-nature-lawsuits/. Accessed 5 Dec 2022.

Christian, David. 2011a. A single historical continuum. *Cliodynamics: The Journal of Quantitative History and Cultural Evolution* 2 (1): 6–26. https://doi.org/10.21237/C7CLIO21203.

———. 2011b. *Maps of time. An introduction to big history*. Bekeley: University of California Press.

Chwałczyk, Franciszek. 2020. Around the Anthropocene in eighty names—considering the urbanocene proposition. *Sustainability* 12 (11): https://doi.org/10.3390/su12114458.

Coen, Deborah R. 2016. Big is a thing of the past: Climate change and methodology in the history of ideas. *Journal of the History of Ideas* 77 (2): 305–321. https://doi.org/10.1353/jhi.2016.0019.

Cohen, Deborah, and Peter Mandler. 2015. The history manifesto: A critique. *American Historical Review* 120 (2): 530–542.

Colebrook, Claire. 2017. We have always been post-anthropocene: The Anthropocene counterfactual. In *Anthropocene feminism*, ed. Richard A. Grusin, 1–20. Minneapolis: University of Minnesota Press.

Costanza, Robert, Lisa J. Graumlich, and Will Steffen, eds. 2007a. *Sustainability or collapse? An integrated history and future of people on earth*. Cambridge, MA, London: MIT Press.

Costanza, Robert, Lisa J. Graumlich, Will Steffen, Carole Crumley, John A. Dearing, Kathy Hibbard, Rik Leemans, Charles L. Redman, and David Schimel. 2007b. Sustainability or collapse: What can we learn from integrating the history of humans and the rest of nature? *Ambio* 36 (7): 522–527. https://doi.org/10.1579/0044-7447(2007)36[522:SOCWCW]2.0.CO;2.

Costanza, Robert, Sander van der Leeuw, Kathy Hibbard, Steve Aulenbach, Simon Brewer, Michael Burek, Sarah E. Cornell, Carole Crumley, John A. Dearing, Carl Folke, Lisa J. Graumlich, Michelle Hegmon, Scott Heckbert, Stephen T. Jackson, Ida Kubiszewski, Vernon Scarborough, Paul Sinclair, Sverker Sörlin, and Will Steffen. 2012. Developing an integrated history and future of people on earth (IHOPE). *Current Opinion in Environmental Sustainability* 4 (1): 106–114. https://doi.org/10.1016/j.cosust.2012.01.010.

Cunha, Daniel. 2015. The geology of the ruling class? *The Anthropocene Review* 2 (3): 262–266. https://doi.org/10.1177/2053019615607069.

Cushman, Gregory T. 2013. *Guano and the opening of the Pacific world. A global ecological history*. Cambridge: Cambridge University Press.

Damianos, Alex und Colin N. Waters. 2018. Minutes. Anthropocene working group (AWG) meeting Max Planck Institute for Chemistry Mainz, Germany. 06/09/2018-07/09/2018. s.l.

Darpö, 2021. Can nature get it right? A study on rights of nature in the European context. Brussels. Accessed 5 Dec 2022.

Daston, Lorraine. 1998. Die Kultur der wissenschaftlichen Objektivität. In *Naturwissenschaft, Geisteswissenschaft, Kulturwissenschaft: Einheit—Gegensatz—Komplementarität?* ed. Lorraine Daston, Otto Gerhard Oexle, and Dieter Simon, 9–40. Göttinger Gespräche zur Geschichtswissenschaft, Bd. 6. Göttingen: Wallstein-Verl.

Demos, T. J. 2017. *Against the Anthropocene: Visual culture and environment today*. Berlin: Sternberg Press.

Deutsche Akademie der Naturforscher Leopoldina e.V.Oncken, Onno, Blanckenburg, Friedhelm von, Wörner, Gerhard and Schlosser, Peter. Erdsystemwissenschaft. Forschung für eine Erde im Wandel. Zukunftsreport Wissenschaft / Deutsche Akademie der Naturforscher Leopoldina e. V.—Nationale Akademie der Wissenschaften. Halle (Saale). Accessed 8 Dec 2022.

Donges, Jonathan F., Wolfgang Lucht, Finn Müller-Hansen, and Will Steffen. 2017. The technosphere in earth system analysis: A coevolutionary perspective. *The Anthropocene Review* 4 (1): 23–33. https://doi.org/10.1177/2053019616676608.

Dukes, Paul. 2011. *Minutes to midnight. History and the Anthropocene era from 1763*. London, New York, Delhi: Anthem Press.

———. 2013. Big history, deep history and the Anthropocene. *History Today* 63 (11): 4–5.

Edwards, Lucy E., Andrew Bauer, Matthew Edgeworth, Erle Ellis, Stanley Finney, Philip L. Gibbard, Jacquelyn L. Gill, Mark Maslin, Dorothy Merritts, William Ruddiman, and Michael J. C. Walker. 2022. The Anthropocene serves science better as an event, rather than an epoch. *Journal of Quaternary Science* 37 (7): 1188. https://doi.org/10.1002/jqs.3475.

Edwards, Paul N. 2017. Knowledge infrastructures for the Anthropocene. *The Anthropocene Review* 4 (1): 34–43. https://doi.org/10.1177/2053019616679854.

Ekström, Anders, and Staffan Bergwik, eds. 2022. *Times of history, times of nature. Temporalization and the limits of modern knowledge. Time and the world.* Vol. Bd. 5. New York, Oxford: Berghahn.

Ellis, Erle C. 2023a. Why I resigned from the Anthropocene Working Group today. https://twitter.com/erleellis/status/1679549881893920772.

———. 2023b. Why I Resigned from the Anthropocene Working Group.

Ellis, Erle C. 2024. The Anthropocene condition: Evolving through social-ecological transformations. *Philosophical Transactions of the Royal Society of London. Series B, Biological Sciences* 379 (1893): 20220255. https://doi.org/10.1098/rstb.2022.0255.

Elverskog, Johan. 2014. Asian studies + Anthropocene. *The Journal of Asian Studies* 73 (4): 963–974.

Emmett, Robert, and Thomas Lekan, eds. 2016. *Whose Anthropocene? Revisiting Dipesh Chakrabarty's "Four Theses"*. München: RCC Perspectives: Transformations in Environment and Society.

Ereshefsky, Marc. 2016. Species. In *Stanford encyclopedia of philosophy*, ed. Edward N. Zalta. Stanford: Metaphysics Research Lab, Stanford University.

Etchart, Linda. 2022. Global governance of the environment, indigenous peoples and the rights of nature. In *Extractive Industries in the Ecuadorian Amazon*, Governance, development, and social inclusion in Latin America Ser. Cham: Springer.

Finney, Stanley C. 2009. International Commission on Stratigraphy. Compiled ICS Subcommission Annual Reports for 2009. s.l. http://www.stratigraphy.org/images/Archive/ICS_AnnReport2009.pdf. Accessed 4 Apr 2019.

Finney, Stanley C., and Lucy E. Edwards. 2016. The "Anthropocene" epoch: Scientific decision or political statement? *GSA Today* 26 (3): 4–10. https://doi.org/10.1130/GSATG270A.1.

Finney, Stanley C., and Philip L. Gibbard. 2023. The humanities are invited to the Anthropocene event but not to the Anthropocene series/epoch: A response to Chvostek (2023). *Journal of Quaternary Science* 38 (4): 461–462. https://doi.org/10.1002/jqs.3520.

Gauger, Jörg-Dieter and Günther Rüther (eds.). 2007. Warum die Geisteswissenschaften Zukunft haben! Ein Beitrag zum Jahr der Geisteswissenschaften, 1. Aufl. Freiburg: Herder.

Geary, Patrick, and Krishna Veeramah. 2016. Mapping European population movement through genomic research. *Medieval Worlds* 4:65–78. https://doi.org/10.1553/medievalworlds_no4_2016s65.

Ghosh, Amitav. 2017. *Die grosse Verblendung. Der Klimawandel als das Undenkbare*. München: Karl Blessing Verlag.

Gibbard, Philip L. 2008. Subcommission on quaternary stratigraphy. Annual Report 2008. Cambridge. http://quaternary.stratigraphy.org/wp-content/uploads/2018/04/SQSAnnual-report-08.doc. Accessed 20 Feb 2019.

———. 2010. Subcommission on quaternary stratigraphy. Annual Report 2010. Cambridge. http://quaternary.stratigraphy.org/wp-content/uploads/2018/04/SQSAnnual-report10.doc. Accessed 20 Feb 2019.

Gibbard, Philip L., and John Lewin. 2016. Partitioning the quaternary. *Quaternary Science Reviews* 151:127–139. https://doi.org/10.1016/j.quascirev.2016.08.033.

Gibbard, Philip L., and Michael J. C. Walker. 2014. The term `Anthropocene` in the context of formal geological classification. In *A sratigraphical basis for the anthropocene?* ed. Colin N. Waters, Jan A. Zalasiewicz, Mark W. Williams, Michael A. Ellis, and Andrea M. Snelling, vol. Bd. 395, 29–37. London: Geological Society of London Special Publications.

Gibbard, Philip L., Michael J. C. Walker, Andrew Bauer, Matthew Edgeworth, Lucy E. Edwards, Erle Ellis, Stanley Finney, Jacquelyn L. Gill, Mark Maslin, Dorothy Merritts, and William Ruddiman. 2022. The Anthropocene as an event, not an epoch. *Journal of Quaternary Science* 37 (3): 395–399. https://doi.org/10.1002/jqs.3416.

Gradstein, Felix M. 2020. Introduction. In *Geologic time scale 2020*, ed. Felix M. Gradstein, James G. Ogg, Mark D. Schmitz, and Gabi M. Ogg, 3–20. Elsevier.
Gradstein, Felix M., James G. Ogg, Alan G. Smith, Wouter Bleeker, and Lucas J. Lourens. 2005. A new geologic time scale, with special reference to Precambrian and Neogene. *Episodes* 27 (2): 83–100.
Graf, Rüdiger. 2012. Zeit und Zeitkonzeptionen in der Zeitgeschichte. s.l. Accessed 4 Apr 2018.
Greenfield, Patrick. 2020. 'Sweet City': The Costa Rica suburb that gave citizenship to bees, plants and trees. *The Guardian*. 29 April 2020. https://www.theguardian.com/environment/2020/apr/29/sweet-city-the-costa-rica-suburb-that-gave-citizenship-to-bees-plants-and-trees-aoe. Accessed 24 July 2023.
Griffiths, Tom. 2002. How many trees make a forest?: Cultural debates about vegetation change in Australia. *Australian Journal of Botany* 50 (4): 375–389. https://doi.org/10.1071/BT01046.
Grunwald, Armin. 2015. Transformative Wissenschaft—eine neue Ordnung im Wissenschaftsbetrieb? *GAIA* 24 (1): 17–20. https://doi.org/10.14512/gaia.24.1.5.
Haff, Peter K. 2014a. Humans and technology in the Anthropocene: Six rules. *The Anthropocene Review* 1 (2): 126–136. https://doi.org/10.1177/2053019614530575.
———. 2014b. Technology as a geological phenomenon: Implications for human well-being. *Geological Society, London, Special Publications* 395 (1): 301–309. https://doi.org/10.1144/SP395.4.
———. 2017. Being human in the Anthropocene. *The Anthropocene Review* 4 (2): 103–109. https://doi.org/10.1177/2053019617700875.
Halfmann, Jost and Johannes Rohbeck (eds.). 2007. Zwei Kulturen der Wissenschaft—revisited, 1. Aufl. Weilerswist: Velbrück Wissenschaft.
Hamilton, Clive. 2016. The theodicy of the "Good Anthropocene". *Environmental Humanities* 7 (1): 233–238. https://doi.org/10.1215/22011919-3616434.
———. 2017. *Defiant earth. The fate of humans in the Anthropocene*. Cambridge, Malden, MA: Polity.
Hamilton, Clive, Christophe Bonneuil, and François Gemenne, eds. 2015. *The Anthropocene and the global environmental crisis. Rethinking modernity in a new epoch*, Routledge environmental humanities series. London: Routledge.
Haraway, Donna Jeanne. 2003. *The companion species manifesto. Dogs, people, and significant otherness*. Chicago: Prickly Paradigm Press.
———. 2010. *Simians, cyborgs, and women. The reinvention of nature*. New York: Routledge.
———. 2015. Anthropocene, Capitalocene, Plantationocene, Chthulucene: Making kin. *Environmental Humanities* 6:159–165.
———. 2016. *Staying with the trouble. Making Kin in the Chthulucene*. Durham: Duke University Press.
Hartley, Daniel. 2016. Anthropocene, Capitalocene, and the problem of culture. In *Anthropocene or Capitalocene?: Nature, history, and the crisis of capitalism*, ed. Jason W. Moore, 154–165. Oakland, CA: PM Press.
Head, Lesley. 2014. Contingencies of the Anthropocene: Lessons from the 'Neolithic'. *The Anthropocene Review* 1 (2): 113–125. https://doi.org/10.1177/2053019614529745.
Head, Martin J., Will Steffen, David Fagerlind, Colin N. Waters, Clément Poirier, Jaia Syvitski, Jan A. Zalasiewicz, Anthony D. Barnosky, Alejandro Cearreta, Catherine Jeandel, Reinhold Leinfelder, John Robert McNeill, Neil L. Rose, Colin Summerhayes, Michael Wagreich, and Jens Zinke. 2021. The great acceleration is real and provides a quantitative basis for the proposed Anthropocene Series/Epoch. *Episodes*1–18. https://doi.org/10.18814/epiiugs/2021/021031.
Head, Martin J., Jan A. Zalasiewicz, Colin N. Waters, Simon D. Turner, Mark W. Williams, Anthony D. Barnosky, Will Steffen, Michael Wagreich, Peter K. Haff, Jaia Syvitski, Reinhold Leinfelder, Francine M. G. McCarthy, Neil L. Rose, Scott L. Wing, Zhisheng An, Alejandro Cearreta, Andrew B. Cundy, Ian J. Fairchild, Yongming Han, A. Juliana, Ivar do Sul, Catherine Jeandel, John Robert McNeill, and Colin P. Summerhayes. 2022. The proposed Anthropocene

epoch/series is underpinned by an extensive array of mid-20 th century stratigraphic event signals. *Journal of Quaternary Science* 37 (7): 1181–1187. https://doi.org/10.1002/jqs.3467.

———. 2023. The Anthropocene is a prospective epoch/series, not a geological event. *Episodes* 46 (2): 229–238. https://doi.org/10.18814/epiiugs/2022/022025.

Hecht, Gabrielle. 2018. Interscalar vehicles for an African Anthropocene: On waste, temporality, and violence. *Cultural Anthropology* 33 (1): 109–141. https://doi.org/10.14506/ca33.1.05.

———. 2023. *Residual governance. How South Africa foretells planetary futures*. Durham: Duke University Press.

Hedberg, Hollis D. 1976. *International stratigraphic guide. A guide to stratigraphic classification, terminology, and procedure*. New York: Wiley.

Hillis, Danny, Rob Seaman, Steve Allen und Jon Giorgini. 2011. Time in the 10.000-year clock. Accessed 23 June 2019.

Hoag, Colin, and Jens-Christian Svenning. 2017. African environmental change from the Pleistocene to the Anthropocene. *Annual Review of Environment and Resources* 42 (1): 27–54. https://doi.org/10.1146/annurev-environ-102016-060653.

Hodder, Ian. 2012. *Entangled. An archaeology of the relationships between humans and things*. Malden: Wiley-Blackwell.

Holmes, Arthur. 1947. The construction of a geological time-scale. *Transactions of the Geological Society of Glasgow* 21 (1): 117–152. https://doi.org/10.1144/transglas.21.1.117.

Horn, Eva, and Hannes Bergthaller. 2020. *The Anthropocene. Key issues for the humanities*. Abingdon, New York: Routledge.

Hornborg, Alf. 2013. Technology as fetish: Marx, Latour, and the cultural foundations of capitalism. *Theory, Culture and Society* 31 (4): 119–140. https://doi.org/10.1177/0263276413488960.

———. 2015. The political ecology of the Technocene: Uncovering ecologically unequal exchange in the world-system. In *The Anthropocene and the global environmental crisis: Rethinking modernity in a new epoch*, ed. Clive Hamilton, Christophe Bonneuil, and François Gemenne, 57–69. Routledge Environmental Humanities Series. London: Routledge.

Hudson, Mark J. 2014. Placing Asia in the Anthropocene: Histories, vulnerabilities, responses. *The Journal of Asian Studies* 73 (4): 941–962.

International Geosphere-Biosphere Program. 2010. Developing an Integrated History and Future of People on Earth (IHOPE): Research Plan. IGBP Report No. 59. Stockholm. Accessed 16 July 2019.

Issberner, Liz-Rejane, and Philippe Lena, eds. 2017. *Brazil in the Anthropocene: Conflicts between predatory development and environmental policies*. London: Routledge.

IUGS. 2024. The Anthropocene. https://www.iugs.org/_files/ugd/f1fc07_40d1a7ed58de458c9f8f24de5e739663.pdf?index=true. Accessed 26 Mar 2024.

Jasanoff, Sheila S. 2020. Ours is the earth: Science and human history in the Anthropocene. *Journal of the Philosophy of History* 14:337–358.

Jordheim, Helge. 2014. Introduction: Multiple times and the work of synchronization. *History and Theory* 53 (4): 498–518.

———. 2017. Europe at different speeds: Asynchronicities and multiple times in European conceptual history. In *Conceptual history in the European space*, ed. Willibald Steinmetz, Michael Freeden, and Javier Fernández-Sebastián, 47–62. New York: Berghahn Books.

———. 2022. Natural histories for the Anthropocene: Koselleck's theories and the possibility of a history of lifetimes. *History and Theory* 61 (3): 391–425. https://doi.org/10.1111/hith.12268.

Jordheim, Helge, and Einar Wigen. 2018. Conceptual synchronisation: From progress to crisis. *Millennium: Journal of International Studies* 46 (3): 421–439. https://doi.org/10.1177/0305829818774781.

Kamel, Lorenzo. 2022. The Anthropocene: History and legacy of the colonial mindset. In *The newsmagazin of the American Historical Association*.

Kistler, Sebastian. 2018. Wie viel Gleichheit ist gerecht? Sozialethische Untersuchungen zu einem nachhaltigen und gerechten Klimaschutz. Dissertation. Beiträge zur sozialwissenschaftlichen Nachhaltigkeitsforschung, Bd. 23. Marburg.

Klingan, Katrin, Jürgen Renn, Bernd Scherer, N. Colin, and Waters und Simon Turner. 2023. *Announcing the proposed GSSP candidate site of the Anthropocene*. Berlin: Lille.

Knauß, Stefan. 2018. Conceptualizing human stewardship in the Anthropocene: The rights of nature in Ecuador, New Zealand and India. *Journal of Agricultural and Environmental Ethics* 31 (6): 703–722. https://doi.org/10.1007/s10806-018-9731-x.

Koselleck, Reinhart, and Hans-Georg Gadamer. 2013. Zeitschichten. In *Studien zur Historik*, vol. 3. Aufl. Frankfurt a. M: Suhrkamp.

Koster, Emlyn. 2023. Public-minded reflections from the Anthropocene working group meeting in Germany. *Episodes* 46 (2): 317–324. https://doi.org/10.18814/epiiugs/2022/022033.

Koyré, Alexandre. 2008. *Von der geschlossenen Welt zum unendlichen Universum*, 2. Aufl. Suhrkamp-Taschenbuch Wissenschaft, Bd. 320. Frankfurt am Main: Suhrkamp.

Latour, Bruno. 2009. Will non-humans be saved? An argument in ecotheology. *Journal of the Royal Anthropological Institute* 15:459–475.

———. 2014a. Agency at the Time of the Anthropocene. *New Literary History* 45 (1): 1–18. https://doi.org/10.1353/nlh.2014.0003.

———. 2014b. How better to register the agency of things. Tanner Lectures, Yale., 26 März 2014. Accessed 22 May 2017.

———. 2014c. Anthropology at the time of the Anthropocene—a personal view of what is to be studied, Washington, DC, 14 December 2014. Accessed 22 May 2017.

———. 2015. Telling friends from foes in the time of the Anthropocene. In *The Anthropocene and the global environmental crisis: Rethinking modernity in a new epoch*, Routledge environmental humanities series, ed. Clive Hamilton, Christophe Bonneuil, and François Gemenne, 145–155. London: Routledge.

———. 2016. Is Geo-logy the new umbrella for all the sciences? Hints for a neo-Humboldtian University, Ithaca, 25 Oktober 2016. Accessed 22 May 2017.

LeCain, Timothy J. 2015. Against the Anthropocene. A neo-materialist perspective. *International Journal of History, Culture and Modernity* 3 (1): 1–28.

———. 2017. *The matter of history. How things create the past*. Cambridge: Cambridge University Press.

Lewis, Simon L., and Mark A. Maslin. 2015a. A transparent framework for defining the Anthropocene epoch. *The Anthropocene Review* 2 (2): 128–146. https://doi.org/10.1177/2053019615588792.

———. 2015b. Defining the Anthropocene. *Nature* 519 (7542): 171–180. https://doi.org/10.1038/nature14258.

Ljungkvist, John, Stefan Barthel, Göran Finnveden, and Sverker Sörlin. 2010. The urban Anthropocene: Lessons for sustainability from the environmental history of Constantinople. In *The urban mind: Cultural and environmental dynamics*, Studies in global archaeology, ed. Paul J.J. Sinclair, Gullög Nordquist, Frands Herschend, and Christian Isendahl, vol. Bd. 15, 367–390. Uppsala: African and comparative archaeology Dept. of Archaeology and Ancient History Uppsala Univ.

Lorimer, Jamie. 2017. The Anthropo-scene: A guide for the perplexed. *Social Studies of Science* 4 (1): 117–142. https://doi.org/10.1177/0306312716671039.

Malm, Andreas, and Alf Hornborg. 2014. The geology of mankind?: A critique of the Anthropocene narrative. *The Anthropocene Review* 1 (1): 62–69. https://doi.org/10.1177/2053019613516291.

Manemann, Jürgen. 2014. *Kritik des Anthropozäns. Plädoyer für eine neue Humanökologie*. Bielefeld, Berlin: Transcript; De Gruyter.

Marquard, Jens. 2019. Worlds apart?: The global south and the Anthropocene. In *The Anthropocene debate and political science*, ed. Thomas Hickmann, Lena Partzsch, Philipp Pattberg, and Sabine Weiland, 200–218. Abingdon, New York: Routledge.

McAfee, Kathleen. 2016. The politics of nature in the anthropocene. In *"Whose Anthropocene? Revisiting Dipesh Chakrabarty's 'Four Theses,'". RCC perspectives: Transformations in environment and society*, vol. 2, 65–72.

McNeill, John Robert. 2019. The industrial revolution and the Anthropocene. In *The Anthropocene as a geological time unit: A guide to the scientific evidence and current debate*, ed. Jan A. Zalasiewicz, Colin N. Waters, Mark W. Williams, and Colin P. Summerhayes, 250–254. Cambridge: Cambridge University Press.

McNeill, John Robert, and Peter Engelke. 2014. *The Great Acceleration. An Environmental History of the Anthropocene since 1945*. Cambridge, MA, London.

Mittelstraß, Jürgen. 1991. Geist, Natur und die Liebe zum Dualismus—Wider den Mythos von zwei Kulturen. In *Glanz und Elend der zwei Kulturen: Über die Verträglichkeit der Natur- und Geisteswissenschaften*, Konstanzer Bibliothek, ed. Helmut Bachmaier and Ernst Peter Fischer, vol. Bd. 16, 9–28. Konstanz: Univ.-Verl. Konstanz.

Moore, Jason W. 2018. The Capitalocene part II: Accumulation by appropriation and the centrality of unpaid work/energy. *The Journal of Peasant Studies* 45 (2): 237–279. https://doi.org/10.1080/03066150.2016.1272587.

Morton, Timothy. 2014. How I learned to stop worrying and love the term Anthropocene. *Cambridge Journal of Postcolonial Literary Inquiry* 1 (2): 257–264. https://doi.org/10.1017/pli.2014.15.

Murphy, Michael A., and Amos Salvador. 1999. International stratigraphic guide—an abridged version. *Episodes* 22 (4): 255–272.

National Assembly, Legislative and Oversight Committee. 2008. Constitution of the Republic of Ecuador.

Nayeri, Kamran Angus Ian. 2016. 'Capitalism in the Web of Life'—a critique. https://climateandcapitalism.com/2016/07/19/capitalism-in-the-web-of-life-a-critique/. Accessed 17 Aug 2020.

Oomen, Jeroen. 2021. *Imagining climate engineering. Dreaming of the designer climate*. Milton: Taylor & Francis Group.

Oppermann, Serpil, and Serenella Iovino, eds. 2017. *Environmental humanities. Voices from the Anthropocene*. London, New York: Rowman & Littlefield International.

Osterhammel, Jürgen. 2017. *Die Flughöhe der Adler. Historische Essays zur globalen Gegenwart*. München: C. H. Beck.

Otter, Chris. 2022. Socializing the technosphere. *Technology and Culture* 63 (4): 953–978.

Otter, Chris, Alison Bashford, John L. Brooke, Fredrik Albritton Jonsson, and Jason M. Kelly. 2018. Roundtable: The Anthropocene in British history. *Journal of British Studies* 57 (3): 568–596. https://doi.org/10.1017/jbr.2018.79.

Pádua, José Augusto. 2017. Brazil in the history of the Anthropocene. In *Brazil in the Anthropocene: Conflicts between predatory development and environmental policies*, ed. Liz-Rejane Issberner and Philippe Lena, 19–40. London: Routledge.

Palsson, Gisli, Bronislaw Szerszynski, Sverker Sörlin, John Marks, Bernard Avril, Carole Crumley, Heide Hackmann, Poul Holm, John Ingram, Alan Kirman, Mercedes Pardo Buendía, and Rifka Weehuizen. 2013. Reconceptualizing the 'Anthropos' in the Anthropocene: Integrating the social sciences and humanities in global environmental change research. *Environmental Science & Policy* 28:3–13. https://doi.org/10.1016/j.envsci.2012.11.004.

Pfister, Christian. 2010. The "1950s Syndrome" and the transition from a slow-going to a rapid loss of global sustainability. In *The turning points of environmental history*, ed. Frank Uekötter, 90–118. Pittsburgh: University of Pittsburgh Press.

Quenet, Grégory. 2017. The Anthropocene and the time of historians. *Annales. Histoire, Sciences Sociales* 72 (2): 165–197. https://doi.org/10.1017/ahsse.2019.9.

Renn, Jürgen. 2020. *The evolution of knowledge. Rethinking science for the Anthropocene*. Princeton, Oxford: Princeton University Press.

República de Colombia, Corte Suprema de Justicia. 2018. STC 4360–2018.

Robin, Libby. 2015. A future beyond numbers. In *Welcome to the Anthropocene: The earth in our hands*, ed. Nina Möllers, Christian Schwägerl, and Helmuth Trischler, 19–24. München: Deutsches Museum.

Robin, Libby, and Will Steffen. 2007. History for the Anthropocene. *History Compass* 5 (5): 1694–1719. https://doi.org/10.1111/j.1478-0542.2007.00459.x.

Robin, Libby, Paul Warde, and Sverker Sörlin, eds. 2013. *The future of nature. Documents of global change*. New Haven: Yale University Press.

Rockström, Johan, Joyeeta Gupta, Dahe Qin, Steven J. Lade, Jesse F. Abrams, Lauren S. Andersen, David I. Armstrong, Xuemei Bai McKay, Govindasamy Bala, Stuart E. Bunn, Daniel Ciobanu, Fabrice DeClerck, Kristie Ebi, Lauren Gifford, Christopher Gordon, Syezlin Hasan, Norichika Kanie, Timothy M. Lenton, Sina Loriani, Diana M. Liverman, Awaz Mohamed, Nebojsa Nakicenovic, David Obura, Daniel Ospina, Klaudia Prodani, Crelis Rammelt, Boris Sakschewski, Joeri Scholtens, Ben Stewart-Koster, Thejna Tharammal, Detlef van Vuuren, Peter H. Verburg, Ricarda Winkelmann, Caroline Zimm, Elena M. Bennett, Stefan Bringezu, Wendy Broadgate, Pamela A. Green, Lei Huang, Lisa Jacobson, Christopher Ndehedehe, Simona Pedde, Juan Rocha, Marten Scheffer, Lena Schulte-Uebbing, Wim de Vries, Cunde Xiao, Xu Chi, Xu Xinwu, Noelia Zafra-Calvo, and Xin Zhang. 2023. Safe and just earth system boundaries. *Nature*. https://doi.org/10.1038/s41586-023-06083-8.

Roelvink, Gerda. 2013. Rethinking species-being in the Anthropocene. *Rethinking Marxism* 25 (1): 52–69. https://doi.org/10.1080/08935696.2012.654700.

Ruccio, David F. Anthropocene—or how the world was remade by capitalism. https://anticap.wordpress.com/2011/03/04/anthropocene%E2%80%94or-how-the-world-was-remade-by-capitalism/. Accessed 1 Aug 2019.

Saito, Kohei. 2017. Marx in the Anthropocene: Value, metabolic rift, and the non-Cartesian dualism. *Zeitschrift für kritische Sozialtheorie und Philosophie* 4:1–2. https://doi.org/10.1515/zksp-2017-0013.

Samida, Stefanie, and Jörg Feuchter. 2016. Why archaeologists, historians and geneticists should work together—and how. *Medieval Worlds* 4:5–21. https://doi.org/10.1553/medievalworlds_no4_2016s5.

Schmidt, Jeremy J., Peter G. Brown, and Christopher J. Orr. 2016. Ethics in the Anthropocene: A research agenda. *The Anthropocene Review* 3 (3): 188–200. https://doi.org/10.1177/2053019616662052.

Schneidewind, Uwe. 2015. Transformative Wissenschaft—Motor für gute Wissenschaft und lebendige Demokratie. *GAIA* 24 (2): 88–91. https://doi.org/10.14512/gaia.24.2.5.

Sideris, Lisa. 2016. Anthropocene convergences: A report from the field. In *Whose Anthropocene?: Revisiting Dipesh Chakrabarty's "Four Theses", RCC perspectives: Transformations in environment and society*, ed. Robert Emmett and Thomas Lekan, vol. 2, 89–96. München.

Sieferle, Rolf Peter. 1982. Der unterirdische Wald. In *Energiekrise und Industrielle Revolution*. München: Beck.

———. 1994. *Epochenwechsel. Die Deutschen an der Schwelle zum 21. Jahrhundert*. Berlin: Propyläen.

Smil, Vaclav. 2015. It's too soon to call this the Anthropocene era. *IEEE Spectrum* 52 (6): 28.

Snow, Charles Percy. 1967. *Die zwei Kulturen. Literarische und naturwissenschaftliche Intelligenz. translated by Grete and Karl-Eberhardt Felten*. Stuttgart: Ernst Klett Verlag.

Soriano, Carles. 2018. On theoretical approaches to the Anthropocene challenge. *The Anthropocene Review* 5 (2): 214–218. https://doi.org/10.1177/2053019618780147.

Sörlin, Sverker. 2018a. Humanities of transformation: From crisis and critique towards the emerging integrative humanities. *Research Evaluation* 14 (1): 287–297. https://doi.org/10.1093/reseval/rvx030.

———. 2018b. Reform and responsibility—the climate of history in times of transformation. *Historisk tidsskrift* 97 (1): 7–23. https://doi.org/10.18261/issn.1504-2944-2018-01-02.

———. 2022. Environmental times: Synchronizing human-earth temporalities from Annales to Anthropocene, 1920s-2020s. In *Times of history, times of nature: Temporalization and the limits of modern knowledge*, Time and the world, ed. Anders Ekström and Staffan Bergwik, vol. Bd. 5, 64–101. New York, Oxford: Berghahn.

Sörlin, Sverker, and Graeme Wynn. 2016. Fire and ice in the academy: The rise of the integrative humanities. *Literary Review of Canada* 24 (6): 14–15.

Steffen, Will. 2021. Introducing the Anthropocene: The human epoch: This article belongs to Ambio's 50th anniversary collection. Theme: Anthropocene. *Ambio* 50 (10): 1784–1787. https://doi.org/10.1007/s13280-020-01489-4.

Steffen, Will, Paul J. Crutzen, and John Robert McNeill. 2007. The Anthropocene: Are humans now overwhelming the great forces of nature? *Ambio* 36 (8): 614–621.

Steffen, Will, Reinhold Leinfelder, Jan A. Zalasiewicz, Colin N. Waters, Mark W. Williams, Colin P. Summerhayes, Anthony D. Barnosky, Alejandro Cearreta, Paul J. Crutzen, Matthew Edgeworth, Erle C. Ellis, Ian J. Fairchild, Agnieszka Galuszka, Jacques Grinevald, Alan Haywood, Juliana Assunção Ivar do Sul, Catherine Jeandel, John Robert McNeill, Eric O. Odada, Naomi Oreskes, Andrew C. Revkin, Daniel D. Richter, James P. M. Syvitski, Davor Vidas, Michael Wagreich, Scott L. Wing, Alexander P. Wolfe, and Hans Joachim Schellnhuber. 2016. Stratigraphic and earth system approaches to defining the Anthropocene. *Earth's Future* 4 (8): 324–345. https://doi.org/10.1002/2016EF000379.

Steffen, Will, and Jamie Morgan. 2021. From the Paris agreement to the Anthropocene and planetary boundaries framework: An interview with Will Steffen. *Globalizations* 18 (7): 1298–1310. https://doi.org/10.1080/14747731.2021.1940070.

Steffen, Will, Åsa Persson, Lisa Deutsch, Jan A. Zalasiewicz, Mark W. Williams, Katherine Richardson, Carole Crumley, Paul J. Crutzen, Carl Folke, Line Gordon, Mario J. Molina, Veerabhadran Ramanathan, Johan Rockström, Marten Scheffer, Hans Joachim Schellnhuber, and Uno Svedin. 2011. The Anthropocene: From global change to planetary stewardship. *Ambio* 40 (7): 739–761. https://doi.org/10.1007/s13280-011-0185-x.

Steffen, Will, Johan Rockström, Katherine Richardson, Timothy M. Lenton, Carl Folke, Diana Liverman, Colin P. Summerhayes, Anthony D. Barnosky, Sarah E. Cornell, Michel Crucifix, Jonathan F. Donges, Ingo Fetzer, Steven J. Lade, Marten Scheffer, Ricarda Winkelmann, and Hans Joachim Schellnhuber. 2018. Trajectories of the earth system in the Anthropocene. *Proceedings of the Academy of Natural Sciences* 115 (33): 8252–8259. https://doi.org/10.1073/pnas.1810141115.

Strohschneider, Peter. 2014. Zur Politik der Transformativen Wissenschaft. In *Die Verfassung des Politischen: Festschrift für Hans Vorländer*, ed. André Brodocz, Dietrich Herrmann, Rainer Schmidt, Daniel Schulz, and Julia Schulze Wessel, 175–192. Wiesbaden: Springer.

Szerszynski, Bronislaw. 2017a. The Anthropocene monument: On relating geological and human time. *European Journal of Social Theory* 4 (1): 111–131. https://doi.org/10.1177/1368431016666087.

———. 2017b. Viewing the technosphere in an interplanetary light. *The Anthropocene Review* 4 (2): 92–102. https://doi.org/10.1177/2053019616670676.

The Long Now Foundation. The 10,000 year clock. http://longnow.org/clock/. Accessed 12 June 2019.

Thomas, Julia Adeney. 2014. History and biology in the Anthropocene: Problems of scale, problems of value. *American Historical Review* 119 (5): 1587–1607.

———. 2017a. The present climate of economics and history. In *Economic development and environmental history in the Anthropocene: Perspectives on Asia and Africa*, ed. Gareth Austin, 291–312. London: Bloomsbury Academic.

———. 2017b. Why do only some places have history?: Japan, the west, and the geography of the past. *Journal of World History* 28 (2): 187–218. https://doi.org/10.1353/jwh.2017.0018.

———, ed. 2022a. *Altered earth. Getting the anthropocene right*. Cambridge: Cambridge University Press.

———. 2022b. Humanities and social sciences: Human stories and the Anthropocene earth system. In *Altered earth: Getting the anthropocene right*, ed. Julia Adeney Thomas, 51–80. Cambridge: Cambridge University Press.

Thomson, Giles, and Peter Newman. 2018. Cities and the Anthropocene: Urban governance for the new era of regenerative cities. *Urban Studies* 7 (1): 1–18. https://doi.org/10.1177/0042098018779769.

Trischler, Helmuth. 2016. The Anthropocene: A challenge for the history of science, technology, and the environment. *NTM* 24 (3): 309–335.

Trischler, Helmuth, and Fabienne Will. 2017. Technosphere, Technocene, and the history of technology. *ICON* 23:1–17.

———. 2019. Die Provokation des Anthropozäns. In *Provokationen der Technikgeschichte: Zum Reflexionszwang historischer Forschung*, ed. Martina Heßler and Heike Weber, 69–105. Paderborn: Ferdinand Schöningh.

Tsing, Anna Lowenhaupt. 2015. *The mushroom at the end of the world. On the possibility of life in capitalist ruins*. Princeton, Oxford: Princeton University Press.

———. 2018. Anthropocene lecture series, Berlin, 27 März 2018.

Turner, Simon, Colin N. Waters, Jan A. Zalasiewicz and Martin J. Head. 2024 What the Anthropocene's critics overlook—and why it really should be a new geological epoch. *The Conversation*. 12 März 2024. https://theconversation.com/what-the-anthropocenes-critics-overlook-and-why-it-really-should-be-a-new-geological-epoch-225493. Accessed 26 Mar 2024.

van der Pluijm, Ben. 2014. Hello Anthropocene, goodbye Holocene. *Earth's Futures* 2 (10): 566–568. https://doi.org/10.1002/2014EF000268.

———. 2018. WTF is the # Meghalayan. Ever thinner (and ever more meaningless) slices for the #geologictimescale. What's next for this silliness? The Kennedyan, the Lennonian, the Mandelian, each with type locality and golden spike? # geology @theIUGS @geosociety [Twitter-Account Ben van der Pluijm]. https://twitter.com/vdpluijm/status/1019587404343148544. Accessed 4 Nov 2019.

Visconti, Guido. 2014. Anthropocene: Another academic invention? *Rendiconti Lincei* 25 (3): 381–392. https://doi.org/10.1007/s12210-014-0317-x.

Vogt, Markus. 2019. Ethik des Wissens. In *Freiheit und Verantwortung der Wissenschaft in Zeiten des Klimawandels*. München: oekom.

Vowinckel, Gerhard. 1994. Verwandtschaft und was die Kultur daraus macht. In *Zwischen Natur und Kultur: Der Mensch in seinen Beziehungen. Beiträge aus dem Funkkolleg "Der Mensch—Anthropologie heute"*, ed. Wulf Schiefenhövel, Christian Vogel, Gerhard Vollmer, and Uwe Opolka, 2nd ed., 32–42. Der Mensch—Anthropologie heute, Beiträge aus dem Funkkolleg 1992-1993 / Red.: Uwe Opolka. Stuttgart: TRIAS Thieme Hippokrates Enke.

Walker, Michael J. C., Philip L. Gibbard and John Lowe. 2015. Comment on "When did the Anthropocene begin? A mid-twentieth century boundary level is stratigraphically optimal" by Jan Zalasiewicz et al. (2015), *Quaternary International*, 282, 196-203. Quaternary International 383: 204–207. doi: https://doi.org/10.1016/j.quaint.2014.11.045.

Walker, Michael J. C., Martin J. Head, Max Berkelhammer, Svante Björck, Hai Cheng, Les C. Cwynar, David Fisher, Vasilios Gkinis, Antony Long, John Lowe, Rewi Newnham, Sune Olander Rasmussen, and Harvey Weiss. 2018. Formal ratification of the subdivision of the Holocene series/ epoch (quaternary system/period): Two new global boundary Stratotype sections and points (GSSPs) and three new stages/ subseries. *Episodes* 41 (4): 213–223. https://doi.org/10.18814/epiiugs/2018/018016.

Walker, Michael J. C., Sigfus Johnsen, Sune Olander Rasmussen, Trevor Popp, Jørgen-Peder Steffensen, Philip L. Gibbard, Wim Hoek, John Lowe, John Andrews, Svante Björck, Les C. Cwynar, Konrad Hughen, Peter Kershaw, Bernd Kromer, Thomas Litt, David J. Lowe, Takeshi Nakagawa, Rewi Newnham, and Jakob Schwander. 2009. Formal definition and dating of the GSSP (global Stratotype section and point) for the base of the Holocene using the Greenland NGRIP ice core, and selected auxiliary records. *Journal of Quaternary Science* 24 (1): 3–17. https://doi.org/10.1002/jqs.1227.

Warde, Paul, Libby Robin, and Sverker Sörlin. 2017. Stratigraphy for the renaissance: Questions of expertise for 'the environment' and 'the Anthropocene'. *The Anthropocene Review* 4 (3): 246–258. https://doi.org/10.1177/2053019617738803.

Waters, Colin N., Simon Turner, Zhisheng An, Anthony Barnosky, Alejandro Cearreta, Andrew Cundy, Ian Fairchild, Barbara Fiałkiewicz-Kozieł, Agnieszka Gałuszka, Jacques Grinevald,

Irka Hajdas, Yongming Han, Martin J. Head, Juliana Ivar do Sul, Catherine Jeandel, Reinhold Leinfelder, Francine McCarthy, John McNeill, Eric Odada, Naomi Oreskes, Clément Poirier, Daniel Richter, Neil Rose, Yoshiki Saito, William Shotyk, Colin Summerhayes, Jaia Syvitski, Davor Vidas, Michael Wagreich, Mark Williams, Scott Wing, Jan Zalasiewicz, Jens Zinke. 2023a. Executive summary: The Anthropocene epoch and Crawfordian age: Proposals by the Anthropocene working group. Submitted to the ICS subcommission on Quaternarty stratigraphy on October 31st, 2023. https://eartharxiv.org/repository/view/6853/. Accessed 27 Mar 2024.

Waters, Colin N., Simon Turner, Jan A. Zalasiewicz, and Martin J. Head, eds. 2023b. *Global boundary Stratotype section and point for the Anthropocene series*. Vol. Bd. 10.

Waters, Colin N., Simon D. Turner, Jan A. Zalasiewicz, and Martin J. Head. 2023c. Candidate sites and other reference sections for the global boundary Stratotype section and point of the Anthropocene series. *The Anthropocene Review* 10 (1): 3–24. https://doi.org/10.1177/20530196221136422.

Waters, Colin N., Mark W. Williams, Jan A. Zalasiewicz, Simon D. Turner, Anthony D. Barnosky, Martin J. Head, Scott L. Wing, Michael Wagreich, Will Steffen, Colin P. Summerhayes, Andrew B. Cundy, Jens Zinke, Barbara Fiałkiewicz-Kozieł, Reinhold Leinfelder, Peter K. Haff, John Robert McNeill, Neil L. Rose, Irka Hajdas, Francine M. G. McCarthy, Alejandro Cearreta, Agnieszka Gałuszka, Jaia Syvitski, Yongming Han, Zhisheng An, Ian J. Fairchild, A. Juliana, and Ivar do Sul and Catherine Jeandel. 2022. Epochs, events and episodes: Marking the geological impact of humans. *Earth-Science Reviews* 234 (Supplement 1): 1–28. https://doi.org/10.1016/j.earscirev.2022.104171.

Waters, Colin N., Jan A. Zalasiewicz, Alex DamianosWaters, N. Colin, Jan A. Zalasiewicz, and Alex Damianos. 2018. *Newsletter of the Anthropocene Working Group. Volume 8: Report of activities 2018*, s.l. http://quaternary.stratigraphy.org/wp-content/uploads/2018/12/Anthropocene-Working-Group-Newsletter-Vol-8.pdf. Accessed 20 Feb 2019.

Watson, Judith, Ted Benton, Kathryn Dean, Pat Devine, Jane Hindley, Richard Kuper, Gordon Peters, Graham Sharp, and Peter Dickens. 2016. Disentangling Capital's Web. *Capitalism Nature Socialism* 27 (2): 103–121. https://doi.org/10.1080/10455752.2016.1178952.

Westermann, Andrea, and Sabine Höhler. 2020. Writing history in the Anthropocene. *Geschichte und Gesellschaft* 46 (4): 579–605. https://doi.org/10.13109/gege.2020.46.4.579.

Wilke, Sabine, and Japhet Johnstone, eds. 2017. *Readings in the Anthropocene. The environmental humanities, German studies, and beyond. New directions in German studies*. Vol. Bd. 19. New York: Bloomsbury Academic & Professional.

Will, Fabienne. 2021. *Evidenz für das Anthropozän. Wissensbildung und Aushandlungsprozesse an der Schnittstelle von Natur-, Geistes- und Sozialwissenschaften*. Göttingen: Vandenhoeck & Ruprecht.

Williams, Rosalind. 1993. Cultural origins and environmental implications of large technological systems. *Science in Context* 6 (2): 377–403.

Witze, Alexandra. 2024a. Geologists reject the Anthropocene as Earth's new epoch—after 15 years of debate. *Nature* 627 (8003): 249 250. https://doi.org/10.1038/d41586-024-00675-8.

———. 2024b. It's final: The Anthropocene is not an epoch, despite protest over vote. *Nature*. https://doi.org/10.1038/d41586-024-00868-1.

Wolff, Eric W. 2014. Ice sheets and the Anthropocene. In *A Sratigraphical Basis for the Anthropocene?* ed. Colin N. Waters, Jan A. Zalasiewicz, Mark W. Williams, Michael A. Ellis, and Andrea M. Snelling, vol. Bd. 395, 255–263. London: Geological Society of London Special Publications.

Yates, Joshua J. 2012. Abundance on trial: The cultural significance of "Sustainability". *The Hedgehog Review* 14 (2): 8–25.

Yusoff, Kathryn. 2018. *A billion black Anthropocenes or none*. Minneapolis: University of Minnesota Press.

Zalasiewicz, Jan A. 2009. Anthropocene Working Group of the Subcommission on Quaternary Stratigraphy (International Commission on Stratigraphy): Newsletter 1, December 2009. http://

quaternary.stratigraphy.org/wp-content/uploads/2018/08/Anthropocene-Working-Group-Newsletter-No1-2009.pdf. Accessed 18 Dec 2019.

Zalasiewicz, Jan A., Colin P. Summerhayes, Martin J. Head, Scott L. Wing, Philip L. Gibbard, and Colin N. Waters. 2019a. Stratigraphy and the geological time scale. In *The Anthropocene as a geological time unit: A guide to the scientific evidence and current debate*, ed. Jan A. Zalasiewicz, Colin N. Waters, Mark W. Williams, and Colin P. Summerhayes, 11–31. Cambridge: Cambridge University Press.

Zalasiewicz, Jan A., Colin N. Waters, Mark W. Williams, and Anthony D. Barnosky. 2019b. Technofossil stratigraphy. In *The Anthropocene as a geological time unit: A guide to the scientific evidence and current debate*, ed. Jan A. Zalasiewicz, Colin N. Waters, Mark W. Williams, and Colin P. Summerhayes, 144–147. Cambridge: Cambridge University Press.

Zalasiewicz, Jan A. und Colin N. Waters. 2012. Anthropocene working group. Report of activities 2011. S.L. http://quaternary.stratigraphy.org/wp-content/uploads/2018/08/Anthropocene-Working-Group-Newsletter-3A.pdf. Accessed 16 Apr 2019.

Zalasiewicz, Jan A., Colin N. Waters, Erle C. Ellis, Martin J. Head, Davor Vidas, Will Steffen, Julia Adeney Thomas, Eva Horn, Colin P. Summerhayes, Reinhold Leinfelder, John Robert McNeill, Agnieszka Gałuszka, Mark W. Williams, Anthony D. Barnosky, B. Daniel de, Philip L. Richter, Jaia Syvitski Gibbard, Catherine Jeandel, Alejandro Cearreta, Andrew B. Cundy, Ian J. Fairchild, Neil L. Rose, A. Juliana, Ivar do Sul, William Shotyk, Simon Turner, Michael Wagreich, and Jens Zinke. 2021. The Anthropocene: Comparing its meaning in geology (Chronostratigraphy) with conceptual approaches arising in other disciplines. *Earth's Future* 9 (3): 1–25. https://doi.org/10.1029/2020EF001896.

Zalasiewicz, Jan A., Colin N. Waters, Mark W. Williams, Anthony D. Barnosky, Alejandro Cearreta, Paul J. Crutzen, Erle C. Ellis, Michael A. Ellis, Ian J. Fairchild, Jacques Grinevald, Peter K. Haff, Irka Hajdas, Reinhold Leinfelder, John Robert McNeill, Eric O. Odada, Clément Poirier, Daniel D. Richter, Will Steffen, Colin P. Summerhayes, James P. M. Syvitski, Davor Vidas, Michael Wagreich, Scott L. Wing, Alexander P. Wolfe, Zhisheng An, and Naomi Oreskes. 2015. When did the Anthropocene begin?: A mid-twentieth century boundary level is stratigraphically optimal. *Quaternary International* 383:196–203. https://doi.org/10.1016/j.quaint.2014.11.045.

Zalasiewicz, Jan A., Mark W. Williams, Alan Smith, Tiffany L. Barry, Angela L. Coe, Paul R. Bown, Patrick Brenchley, David Cantrill, Andrew S. Gale, Philip L. Gibbard, F. John Gregory, Mark W. Hounslow, Andrew C. Kerr, Paul Pearson, Robert Knox, John H. Powell, Colin N. Waters, John Marshall, Michael Oates, Peter F. Rawson, and Philip Stone. 2008. Are we now living in the Anthropocene? *GSA Today* 18 (2): 4–8. https://doi.org/10.1130/GSAT01802A.1.

Zhong, Raymond. 2023. The human age has a new symbol. It's a record of bomb tests and Fossil fuels. *The New York Times*. 11 July 2023. https://www.nytimes.com/2023/07/11/climate/anthropocene-epoch-crawford-lake.html. Accessed 27 Mar 2024.

Chapter 5
Outreach of the AWG in the Media: The Anthroposcene, the Social Sciences, and the Creative Arts

While the term "Anthroposcene" might be considered flippant, the idea has been successfully used to popularize the Anthropocene notion as both a scientific and a cultural phenomenon. Two major multimedia exhibitions in Germany, "The Anthropocene Project" at the museum Haus der Kulturen der Welt (2022) in Berlin and "Welcome to the Anthropocene: The Earth in our hands" in the *Deutsches Museum* (Möllers et al. 2015) in Munich, are outstanding examples of how the AWG embraced the arts and how the arts embraced the Anthropocene. It is well documented that HKW worked closely with the Anthropocene Working Group, going beyond providing meeting places for its early deliberations and extending to large sums of money to fund its Outreach strategies (see Damianos 2021). Damianos's work is particularly interesting as he was an intern for the AWG in 2018, providing help for the AWG website.

This chapter also shows how the Outreach of the AWG provoked hostile comments from some members of the geological community, who were little used to such exposures.[1] It is interesting to compare this contemporary manifestation of culture wars around the Anthropocene with Paul Forman's classic research on culture wars among mathematicians and physicists in the Weimar Republic (see Forman 1971).

This chapter is authored by Leslie Sklair.

---

[1] I want to thank my co-authors of this volume, Martin Bohle, Boris Holzer, and Fabienne Will, for their valuable input and suggestions.

M. Bohle et al., *The Anthropocene Working Group and the Global Debate Around a New Geological Epoch*, Emerging Globalities and Civilizational Perspectives, https://doi.org/10.1007/978-3-031-85175-9_5

5.1 The Right Place and the Right Time

The early proponents of the Anthropocene notion were fortunate to find themselves in the right place at the time. As is well documented, the limnologist Eugene Stoermer had begun to use the term "Anthropocene" informally in the 1980s, and the atmospheric chemist Paul Crutzen (joint Nobel Prize winner in 1995 for research on the ozone layer) is credited with popularizing the term, notably through a short statement in the journal *Nature* in 2002. The combination of Nobel Laureate and *Nature* was undoubtedly a good start in putting the Anthropocene notion into the scientific and public domains. Crutzen's role in creating the AWG in 2009 and its challenging history since then have been exhaustively documented (see Zalasiewicz et al. 2019; Benner et al. 2021; Damianos 2021). The fact that the AWG was created securely within the official apparatus of the geological establishment facilitated access to the most prestigious scholarly publications, notably the *Philosophical Transactions of the Royal Society* (London), *Proceedings of the National Academy of Science* (USA), and of course, *Nature* (widely recognized as the most prestigious science journal in the world). Between 2010 and 2020, the AWG was referenced 28 times in *Nature*, in research articles, news items, and editorials.

In the 2010s, several new English-language journals focused on the Anthropocene began to be published, most importantly *Anthropocene*; *Elementa: Science of the Anthropocene*; both from 2013, and *Anthropocene Review* (2014). Oldfield et al. (2013) provide an interesting rationale for the *Anthropocene Review*, yet another Anthropocene journal, highlighting "transdisciplinarity," a theme that we find in much Anthropocene literature. In addition, the journal *Anthropocene Science: An International Journal for Addressing Human Impact on the Resilience of the Planet Earth* published by Springer Nature—Singapore, has appeared. Also of interest is *Anthropocene: the Online Poetry Journal.* In 2014, the US-based *Conservation Magazine* morphed into the *Anthropocene Magazine.* According to the website, "Our goal is not to make people feel better; nor is it to scare them into paralysis. Rather, we are forging a sophisticated middle ground: evidence-based journalism that puts the best science and innovations into the hands of those who can do the most with them. In other words, we are balancing urgency with efficacy." This new publication targets a more general audience and offers a sponsorship kit. In its first issue, the editor writes: "There's been an ongoing debate in environmental circles about whether the Anthropocene is good or bad. Our take on the question is that it is inherently neither. It simply *is*: it is the rather extraordinary time in history in which we find ourselves. And for good or ill, we are now in the driver's seat. So let's start talking about where we want to go." The editorial highlights the contribution of the former *New York Times* columnist Andrew Revkin, a member of the AWG until 2016 and probably the most influential proponent in the USA of the "good Anthropocene," and also references Elizabeth Kolbert's support of Hamilton (see Sklair 2021, p. 61ff.).

The informative *Anthropocene Magazine—Newsletters* began to publish in 2021, mostly good news stories, inviting its readers to expand its audience with "Know friends or colleagues who might be interested in this newsletter." *Anthropocene Magazine* is supported by the organization Future Earth under the controversial slogan "Welcome to the Anthropocene". This slogan had been picked up by mass media all over the world (most famously or infamously according to one's attitude) in June 2012, when the United Nations Rio + 20 summit on sustainable development was opened with a short film "Welcome to the Anthropocene"—a film about the state of the planet." *The Economist* had used the slogan as a headline in its special issue on the Anthropocene published on May 28, 2011; and in 2014, the Deutsches Museum in Munich opened the multimedia exhibition entitled "Welcome to the Anthropocene: The Earth in Our Hands" (Möllers et al. 2015). For a critical account of this exhibition, see Jørgensen and Jørgensen (2016).

The Australian ethicist Clive Hamilton reacted strongly against "Welcome to the Anthropocene," and he was not alone. His powerful critique was reported in media all over the world. For example, in *The Tribune* (India) and *The Belfast Telegraph* (both in 2016), we find articles on the Anthropocene citing him: "According to Clive Hamilton, some scientists even write: 'Welcome to the Anthropocene adding, 'At first, I thought they were being ironic, but now I see they are not. And that's scary. The idea of the Anthropocene is not welcoming. It should frighten us. And scientists should present it as such" (see also Hamilton 2017, pp. 21-26). While it seems obvious that many who say or write "Welcome to the Anthropocene" are doing so ironically, the irony might be lost on some audiences, and the slogan could easily be seen to demean the science. These varying public representations of the Anthropocene clearly created a dilemma for the AWG in the sense that the rhetoric around the Anthropocene notion vacillated around the optimistic, often neutralizing and/or reassuring language of the "good" Anthropocene and the alarmist, the existential threat to human survival, hinted at by Crutzen in his 2002 piece in *Nature,* and more robustly by Hamilton, Kolbert, and others. Crutzen himself, as coauthor of an article by Steffen and other AWG members in 2011, had argued: "The ultimate drivers of the Anthropocene, on the other hand [compared with the reception of Darwin], if they continue unabated through this century, may well threaten the viability of contemporary civilization and perhaps even the future existence of *Homo sapiens*" (Steffen et al. 2011, p. 862). While the extinction of species in terms of biodiversity is regularly discussed by AWG members, discussion around the extinction of humanity is generally avoided. Barnosky's much-cited multi-authored paper (2011) entitled "Has the earth's sixth mass extinction already arrived?" does not address the issue that humanity itself may be a species at risk. See Ceballos et al. (2017) and Nick Bostrom's Home Page for notable exceptions.

5.2 Public Understanding of Science

In the decades around the new Millennium, debates and research on various aspects of the public understanding of science and public engagement with science were well established and rapidly expanding (see, for example, Friedman et al. 1999;

Brossard 2013; Bucchi and Trench 2014; Marcinkowski et al. 2014; Carver 2014; Peters and Dunwoody 2016; Akin and Landrum 2017; Jamieson et al. 2017; Hara et al. 2019). Of special interest is Dunwoody (1999), pre-Anthropocene but of direct relevance for reporting "uncertainty in science." This was also a period of rapid expansion in the number of science journals in general and also the rise of a new phenomenon, namely the internationalization of Chinese English-Language humanities and social science, and science and technology journals (see respectively Huang and Miao 2021; and Wang et al. 2018). The first use of the term "Anthropocene" in China is attributed to the Chinese geologist, Liu Dongsheng, in 2004 (see Weigelin-Schwiedrzik 2018, p. 19).

There were, of course, by this time, also ample opportunities to popularize (or debunk) scientific ideas provided by the internet, notably via Facebook, Google, Wikipedia, Twitter, and even the Reddit Journal of Science (see Collins et al. 2016; Darling et al. 2013; Hara et al. 2019; Pflugfelder and Mahmou-Werndli 2021). Also relevant in this connection is the book *Science Fictions* (Ritchie 2020). While it does not deal with Earth or other natural sciences, its critique of scholarly methodologies and the journal publications industry is worth pondering.

It is clear that the members of the AWG were well aware of the importance of explaining the Anthropocene not only for geologists and Earth System scientists in general but also for wider scientific communities and other possible allies in its quest for acceptance of the term "Anthropocene" in the official geological nomenclature, as well as in the public realm. A case in point is the outreach of legal scholar Davor Vidas, a specialist on the law of the sea who became a founding member of the AWG (see Zalasiewicz et al. 2019: Section 1.4.3.1). Correspondence between Vidas and Zalasiewicz is available from the AWG Newsletter of 2010. Incidentally, Professor Vidas may be credited for organizing the first ever conference that connected globalization and the Anthropocene (see (Vidas and Schei 2013)).

5.3 AWG Outreach via the AWG Newsletter (2009–2022)

Around 2012, it became clear that AWG members not only participated in scientific matters but also started to take an increasing interest in public and media events. From then on, AWG newsletters published details of various outreach events. For example, we learn that The Duke University Perkins-Bostock Library will show an exhibit, "Welcome to the Anthropocene?" (an interesting choice of title) from July to October 2013. "The exhibit grows from a diverse group of scientists, humanities scholars, and artists at Duke who meet to read stratigraphy, global change science, critical essays, and philosophy. The exhibit includes displays on the pop-cultural concept of the Anthropocene and examines four proposed dates of origin as well as the contrasting public narratives that result from each of the four dates of origin. A reading and cinema list is included." In 2012, the Chairman's Column concluded on a very upbeat note: "This last year has seen continuation of extraordinary interest in the Anthropocene concept, from both within the Earth sciences (writ large) and beyond it. There are museum exhibitions, in progress and planned, the creation of dedicated journals, radio programmes, and the publication of a range of articles, both technical and general. It is clear that the term is being found widely useful. Time will tell whether this widespread interest will continue. But, for the time being, putting all else aside, it is good to see that the aims and working practices of stratigraphy – long regarded as amongst the most arcane areas within geology – are also receiving wide attention." Also announced was the news of a new journal to be published by Sage: *The Anthropocene Review.* This was identified as a transdisciplinary journal, bringing together articles on all aspects of research pertaining to the Anthropocene, earth and environmental sciences, social sciences, material sciences, and humanities. "High impact research articles, reviews and brief 'perspective' articles are especially welcome." Another new journal *Anthropocene* is also announced, published by Elsevier. Reporting of scientific publications and Anthropocene-related conferences and media events by AWG members increased gradually at this time.

Another significant feature of the Newsletters appears to be the New Members Forum, where scholars and creative artists are invited to showcase their Anthropocene-related activities. A wide variety of people from diverse disciplines and worldwide enthusiastically took up the invitation.

As the 2010s progressed, AWG newsletters offered further evidence of the AWG members' efforts to spread the word about the Anthropocene notion to various audiences. It seems clear from these AWG newsletters that what can be understood as Outreach became more central to the AWG—communicating various ways in which the AWG took the matter of reaching out to both other members of the geological and Earth Sciences communities and the publics that were not strictly science-based.

As noted earlier, the AWG worked closely with the "Haus der Kulturen der Welt" in Berlin (Haus der Kulturen der Welt 2022), which sponsored the first in-person of several meetings of the AWG in the Museum, reported in the Newsletter in 2014. The cover of this issue of the Newsletter featured a picture of the New York Night Cityscape, a revealing choice. This meeting in 2014 was regarded as a considerable success. It resulted in widespread publicity (including in *Nature*, with a feature, and an editorial and inclusion in their annual roundup of "highlights of the year": and in the journal *Science* as well as in *The Economist magazine* and on the BBC, and in the *National Geographic*). With some pride, it is reported that "The term 'Anthropocene' was included in the title of the annual Geological Society of America meeting in 2011 and was a major theme at the 2011 Nobel Laureate Symposium in Stockholm, being a key concept in the Stockholm Memorandum, published after the meeting in the journal *Ambio*" (3rd Nobel Laureate Symposium on Global Sustainabilty 2011). Perhaps the most significant achievement reported in the Newsletter at this time was that "[i]n 2016 seven prominent members of the AWG—Erle Ellis, John McNeill, Eric Odada, Andrew Revkin, Will Steffen, Davor Vidas and Jan Zalasiewicz—were interviewed in the feature documentary Anthropocene which showed on campuses and at film festivals worldwide and helped the term to gain public attention. The documentary was the first feature-length film about the new epoch, according to entries at imdb.com and was described by Earth.com as one of the top ten documentaries to help raise environmental awareness. While the seven AWG members formed a broad consensus about the Anthropocene's history and the term's significance, they took contrasting views when invited by director Steve Bradshaw to consider the Anthropocene either as a tragedy—with extinctions and upheavals—or as a dark comedy." This paragraph comes from an entry on the Anthropocene in Wikipedia (2024).

Since 2012–2013, the Newsletter has contained a section labeled MEDIA, beginning with an item from *Süddeutsche Zeitung* as follows: "This significant German newspaper ran an article on 9 March 2012 on the Anthropocene (in German) quoting the work of Paul Crutzen, Erle Ellis and Jan Zalasiewicz."[2] The Newsletter also referenced newspaper and magazine articles involving AWG members in French, as well as a major four-part series on "The Age We Made," broadcast in November 2012 as part of BBC World Service's *Discovery* series and later on BBC Radio 4's *Frontiers*, including AWG members.

Another significant event at this time was the AWG inaugural group meeting of this period, hosted by the HKW Exchanges initiative, described as follows: "Exchanges initiates a set of dialogues between members of the Anthropocene Working Group and social scientists, thinkers, and artists, a serial thread of conversations that draws from a vast range of expertise, disciplines, and practices. It aims to reflect on the 'what' and 'how' of knowledge-articulation in the world, taking the material Earth and its combined history with the human as a starting point and combining it with an idea of the academic as a concerned citizen'. Each of these dialogues addresses the concrete range of hands-on practices that emerge from epistemic infrastructures and worldviews in place, engaging with research methods in the lab" or field, at the desk or in the studio. These dialogues ask how such varied practitioners from the sciences, humanities, arts, and activism use their 'everyday' matters, how their diverse practices are affected and what they effect. What informs their concepts, and what are the differences in their respective terminologies? How do these transform their encounters with other knowledge-forms? And, how can such movements of flux between and across specificities of expertise configure possible pathways for re-establishing modes of collaboration?" Evidence that this agenda was widely accepted by Anthropocene scholars of all types is provided by the fast-growing Anthroposcene-related literature along these lines. Two interesting examples among hundreds are the collections edited by Jagodzinski (2018), and Dürbeck and Hüpkes (2020).

The Exchanges initiative strongly suggests that the interdisciplinarity that was fast becoming a hallmark of Anthropocene studies was something more than the usual interdisciplinarity that we find across many scientific disciplines but was, in fact, being mobilized as a technique of Anthropocene Outreach for all forms of knowledge production and critique.

[2] For a more detailed summary of the Anthropocene in the German media, see Holzer and Sklair (2021, pp. 168–173).

In 2019, Cambridge University Press published a book on the Anthropocene by AWG members (Zalasiewicz et al. 2019). The blurb informs us: "The book set out to provide a summary of the evidence of the Anthropocene assembled to date, much of it published or research underway by members of the Working Group but also summarizing the growing body of wider Anthropocene literature … rigorously laying out technical geological evidence … it is also designed to be coherent and accessible to the anticipated wide audience." … *the rigorously presented stratigraphic evidence will be presented in a manner accessible to people from a broad range of educated backgrounds. The level will be generally that of the more journalistic or general review sections of Nature or Science*" (italics added).

Notably, the AWG, along with its supporters at the HKW, was able to make skilful connections between the hard science of the Anthropocene and what is designated as "beyond the confines of chronostratigraphy." The Newsletter explains: "AWG members have been involved in the Anthropocene's continuing wider reach, through such means as Berlin's HKW Anthropocene Curriculum project, the Burtynsky film *Anthropocene*" (discussed in detail below). Also notable were Erle Ellis's (2018) Anthropocene book (in the prestigious OUP short introduction series) and Peter Haff's blog "Being Human, in the Anthropocene" (Haff 2024). The year 2018 also saw the Anthropocene as a framing concept in the IPCC Special Report on Global Warming and the WWF Living Planet Report of 2018. "These all remind us, if reminder was needed, that our particular corner of stratigraphy is not just a thing of the past." Jan Zalasiewicz concludes his report as chairman of the AWG with this highly optimistic statement: "The AWG has, collectively, brought something quite new to geology, and beyond. When we started in 2009, literature on the Anthropocene, as Davor Vidas has noted, could be skim-read in an afternoon, with 64 Google hits on 27 August 2009 for the term Anthropocene; as I write, Google shows around 5 million hits. It has been quite an adventure to be caught up in this phenomenon: I look forward to continuing this adventure with you all." In 2019, Jan Zalasiewicz resigned as chair of the AWG, to become Chair of the Subcommission on Quaternary Stratigraphy.

In 2018, Bernd Scherer, Director of the HKW, announced that funding had been secured for the cultural-scientific project "Earth Indices" as part of HKW's long-term investigation on the Anthropocene. This would allow the HKW to collaborate even closer with the Anthropocene Working Group and contribute to the GSSP analysis programme, which will be presented to the public at HKW in the summer of 2022. Also reported was that the Anthropocene was "accepted" by the UNESCO board/IGCP committee. This project is one of 18 new projects to join the International Geoscience Programme in 2021, and there are many more involving UNEP and the EU. All through this account of successes for the Anthropocene notion is the implicit idea that the scientists, creative artists, and others from the HKW are working hand in hand.

Another AWG/HKW joint initiative, "Unearthing the Present," was reported in the Newsletter 2022 [p. 3] as follows: "As ever, HKW were wonderful hosts, and their excellent coordination ensured that the meeting proved a great success. In addition to two days of presentations and discussion around the advanced progress of the 12 candidate GSSP sites, many of the principal investigators were able to be involved with public discussions about their work through a series of live 'core readings' … an outstanding feature of all the research teams has been their positive attitude to disseminating their work outside of the geological community and giving their time to many media requests."

The Newsletter also references the digital publication *Anthropogenic Markers: Stratigraphy and Context* (Edited by Christoph Rosol and Giulia Rispoli 2017). This "explores and contours some of the historical contexts, epistemic settings, and conceptual interventions around the ongoing work of Anthropocene geology. In seven thematic dossiers, contributions from geochemistry and paleobiology, history and science studies, artistic research, archaeology, literary studies, and anthropology reflect on selected material markers of human impacts on earthly strata. Together, they form a unique interdisciplinary conversation across positions and perspectives, helping to map out and expand upon the many ways of tracing the evidence that defines the onset of a human-dominated and crisis-laden epoch." Again, we can discern elements of what can be seen as interdisciplinarity outreach on a scale that is very unusual in scientific discourse.

5.4 Bibliographic Evidence of AWG Outreach Across Multiple Spheres of Scholarship

In 2021, Springer and the Max Planck Institute published a volume to commemorate the work of Paul Crutzen (he died before the volume was published). It was focused on his seminal contribution to the study of the Anthropocene (see Benner et al. 2021). Chapter 22 of the book, by Hans Brauch, was entitled *Bibliometric Analysis of the Anthropocene Literature (May 2000–January 2021)*. The chapter consists of a remarkable listing of daily snapshots to illustrate the data in the *Web of Science* and *Scopus* databases on the topic "The Anthropocene Concept in the Natural and Social Sciences." Brauch's chapter stretches from page 307 to page 438. The chapter in the table of contents has a slightly different title, "The Anthropocene Concept in the Natural and Social Sciences, the Humanities and Law – A Bibliometric Analysis and a Qualitative Interpretation (2000–2020)." The expansion of the title seems significant to me in terms of outreach.

In a series of tables and commentaries, Brauch presents a vivid portrait of what might be considered the Outreach of the AWG across almost all fields of natural sciences and a cornucopia of other scientific disciplines and the creative arts. Listed are thousands of publications not only in the obvious disciplines in and around geology and the environmental humanities but also, for example, in Health professions, Public Administration, Arts, Humanities, Literature, Business, Management, Accounting. Engineering, Government, law, Economics, Econometrics, Finance, Architecture, Genetics, Philosophy, Education, Sociology, Communication,

Physics, Astronomy, Cultural Studies, Psychology, Mathematics, Decision Sciences, Heredity, International Relations, Film, Radio, Television, Neuroscience, Linguistics, Nursing, Veterinary, and Religion. More tables present Anthropocene texts by countries and languages in the *Web of Science* and in *Scopus* (2001–2021). The countries are too numerous to list here. It suffices to say that there are Anthropocene texts from most members of the UN, as well as most major and many minor languages. These texts include hundreds of books and book chapters, reviews, 139 Book Series, 1584 Editorials, and thousands of Conference Proceedings. Anthropocene references appear in over 113 Poetry, 215 Art Exhibits, over 100 Biographical Items, 117 Fiction, Creative prose, 118 Film Review, 119 Music Performance Reviews, and much more (see Brauch 2021).

5.5 The AWG in the Media

The Anthropocene Media Project (AMP) set out to discover the extent and content of Anthropocene-related coverage in print media worldwide in all the major (and some minority) languages. Starting from the year 2000, when the term "Anthropocene" was first identified in the print media, until the end of 2017, over 4000 items containing the term "Anthropocene" were accessed (online) from about 2000 newspapers and magazines in 140 countries. The Anthropocene Working Group was referenced about 200 times in these media searches, and Jan Zalasiewicz, who led the AWG, was referenced about 240 times. Many of his colleagues in the AWG also appeared in newspapers and magazines around the world (see Sklair 2021, *passim*). Media interest seems to have peaked in 2016, largely stimulated by the campaign to bring the idea of the "Anthropocene" into the vocabulary of geologists, and the general public led by the AWG at the 35th International Geological Congress held in Cape Town (see Chap. 2) in that year. Despite the best efforts of the AWG, much of the extensive reporting of the Anthropocene in Cape Town gave the misleading impression that the scientists had come to a consensus that the "Anthropocene" had arrived, for better or worse. However, the geological community saw the idea as speculative (see Chap. 2), and its place in the geological nomenclature was far from settled. This perception is a dilemma that the AWG is still grappling with, and given the latest developments, namely the rejection of the ratification by the Subcommission on Quaternary Stratigraphy in March 2024 (see Chap. 6), will continue to grapple with in the future, increasingly so as social scientists, environmental humanities scholars, and creative artists seek to enlarge the possibilities of Anthropocene (or as some would argue, Anthroposcene). It is important to note that climate change is often used in the media and scholarly publications as a metonym for the Anthropocene (substituting the part for the whole). If these were factored into this chapter, the volume of media coverage would be considerably higher.

Irrespective of these issues of interpretation, it is clear that the AWG, in general, and that Jan Zalasiewicz, in particular, played the leading role in the popularization of the Anthropocene as a global phenomenon and successfully disseminated the general idea to a global audience in a way and to the extent that few other scientific concepts had been popularized before. For example, the following media platforms brought the AWG to the attention of their readers and there are many more: *Global Times, China Daily* (Asian edition), *Gansu Daily*; *Xinhua*; *Science Magazine*; *Vesmír; Outlook*; *The Age; The Conversation; Hobart Mercury*; *Die Presse; Daily Star Bangladesh; La Tercera; Slobodna Dalmacija; Diario Libre; Granma (*borrowed from the BBC); *Gulf News; UusiSuomi* (Finland); *Le Télégramme* (Brittany); *Le Figaro; Frankfurter Allgemeine Zeitung; Focus; Jamaica Gleaner*; *Japan Times; Belfast Telegraph; New Zealand Herald; Newsweek Polska; Público; Gulf Times; Straits Times; Pravda; Daily Mail; Netwerk24; La Razon; El Mundo; La Vanguardia; Aftonbladet; Svenska Dagbladet; Apple Daily; Bangkok Post; Daily Express; The Economist; Daily Telegraph; Utica Observer-Dispatch (Utica, New York); Scranton Times-Tribune; Anchorage Daily News; San Diego Union Tribune; Fox News; Albuquerque Journal; Austin Statesman; Jewish Journal LA; Christian Post; Scientific American; El Nuevo Herald (Doral, Florida); Stuff New Zealand* and *The Guardian.*

The *Guardian* cited the AWG several times; most famously in an article by Kate Raworth (2014) "Must the Anthropocene be a Manthropocene?", a comment on AWG membership. She points out: "But if this first Berlin meeting of the Anthropocene Working Group (AWG) was anything to go by, the matter might as well be settled now: just call it the Manthropocene. Of the 29 scientists the working group listed on its website as members at the time of the meeting, only one was a woman." Of course, every mention of the Manthropocene was a mention of the Anthropocene, Outreach of a sort. As the analysis of AWG membership in Chap. 2 shows, female participation at the AWG increased over time.

However superfluous the list of newspapers and magazines aforementioned might seem, it does provide solid evidence of the success of the AWG in spreading the word and the message of the Anthropocene all over the world to a wide variety of publics via global and local media. This outreach of the AWG represents one of the most successful campaigns for popularizing a new, albeit controversial, concept in the history of science.

Of course, not all social scientists and others who wrote about the Anthropocene were positive about the ideological implications of the term. Those who wished to replace this opposition with the idea of the "Capitalocene" most clearly expressed this opposition, drawing attention to the obvious truth that not all humans were equally responsible for the existential damage. The Global North, capitalism, and colonialism are identified as the main culprits. In his magisterial study on Fossil Capital, the Swedish Marxist scholar Andreas Malm made the connection between the rise of steam power and the roots of global warming, the causes of the so-called "Anthropocene," more accurately labeled the "Capitalocene" according to the critics (see Malm 2016). Jason Moore's edited book popularized the term Capitalocene, attracting media coverage (see Moore 2016). Somewhat diluting the impact of the new term, about 100 other "scenes" followed, in some respects echoing the idea of the "Anthroposcene." Donna Haraway is another scholar regularly cited as a critic of “Anthropocene” and a proponent of the Capitalocene. In her contribution to Moore’s edited collection, she levels seven objections to the Anthropocene concept. However, she concludes: “in so far as the Capitalocene is told in the idiom of fundamentalist Marxism, with all its trappings of Modernity, Progress, and History, that term is subject to the same or fiercer criticism” (in Moore 2016, pp. 52-53).

The view that the term “Anthropocene” distorted the realities of human history obviously resonated with many people who were increasingly dismayed that the proponents of the geological Anthropocene rarely left their ivory towers. This was, to some extent, justified. Very little of the Anthropocene literature in the early years paid much attention to the issues that the Capitalocene and other “cenes” raised. This began to change, albeit slowly, in the 2010s. In 2018, two British scholars published a book aimed at the mass market that engaged with the science, the history, and the Capitalocene. Arguing that “calling this the Capitalocene, is in our view, not correct because the Anthropocene will last so far into the future,” millions of years, probably far outlasting capitalism (Lewis and Maslin 2018, p. 444, note 21). Perhaps, we can see the Capitalocene in terms of “negative/positive Outreach” insofar as every reference to the Capitalocene is also a reference to the Anthropocene.

An interesting article reviewing six books on the Anthropocene published between 2013 and 2021 shows that historians had been writing about the Anthropocene for some years (Chaplin 2023). The obvious reference here is the work of Chakrabarty (2009, 2021). Where we might ask, were historians getting ideas that the Anthropocene might be worth finding more about? The article not only references the AWG but actually engages with its work.

The fact that the AWG was actively monitoring media interest in the Anthropocene is clear from the AWG newsletter. As noted earlier, while much of the Newsletters were taken up by information about GSSP sites and other technical issues, the focus in its latter years appears to be on AWG’s success in popularizing the term “Anthropocene.” For example, in 2020, a color table was printed listing large numbers of conferences, organizations, meetings, lectures, and other forms of communication with any connection to the Anthropocene. The newsletter reported that the number of media events increased about 10-fold between 2018 and 2022 (for a detailed analysis of Anthropocene coverage in the global media, see Sklair 2021, *passim*). These numbers indicate a much-increased activity. The increasing interest in media coverage suggests that AWG members cared about scientific exchange and public visibility. However, through the newsletter, it appears as if not all members of the AWG cared much about working with the media. However, those who were enthusiastic about working with the media clearly were very successful, as evidenced by the range of newspapers from all over the world that published Anthropocene-related news items, often reproducing interviews with AWG members. Notable in this regard was the press information made public by Leicester University (the home campus of Jan Zalasiewicz) on the occasion of the Cape Town International Geological Conference in 2016. The success of those who were campaigning for the Anthropocene notion in bringing their message to the media received grudging support from an influential geology notable, a leading Anthropocene sceptic, Stan Finney, Chair of the International Commission on

Stratigraphy (ICS) from 2008 to 2016. Finney is quoted: "There's this huge, huge media tsunami wave in both scientific publications and the media [around the Anthropocene]" (see Damianos 2021, pp. 122-123.) The final issue of the AWG Newsletter in 2023 ended with the statement: "Although, as a working group we are moving towards the final phase of our remit of producing a formal proposal, we will endeavor to continue researching diverse aspects of the Anthropocene concept and investigate options for continuation of such stimulating debate and research beyond the eventual disbanding of the AWG. We look forward to working with you on this pivotal year in the AWG's history." Even if the scientific aims of the AWG were yet to be decided, there can be no doubt that its broader aim was to ensure that the Anthropocene notion was securely placed in the public realm and that it was on the path to global recognition.

5.6 The Aims of the AWG

In 2019, Jan Zalasiewicz and other members of the AWG published *The Anthropocene as a Geological Time Unit: A Guide to the Scientific Evidence and Current Debate* (Zalasiewicz et al. 2019). This book may be taken as an authoritative statement of the aims of the AWG cleverly combining the geological science and the open-ended idea of a "Current Debate" around the Anthropocene. A review of the book by John Matthews in the journal *Holocene* begins: "The Anthropocene as a Geological Time Unit presents both a justification of the stratigraphical approach used by geologists and a summary of the evidence collected by the Anthropocene Working Group (AWG). This group was set up in 2009 under the auspices of the International Commission on Stratigraphy of the International Union of Geological Sciences to investigate the case for the Anthropocene as a formal geological time unit. *It is not, however, concerned with non-geological interpretations of the Anthropocene, including those of archaeology, the social sciences and the humanities* (italics added). Principles of stratigraphy and the utility of a stratigraphic approach are explained and exemplified with reference to several other stratigraphic boundaries selected from earlier in the geological record" (Matthews 2019, p. 368). Matthews' statement that the AWG "is not, however, concerned with non-geological interpretations of the Anthropocene, including those of archaeology, the social sciences and the humanities" is misleading with respect to both the book under review and to the activities of the members of the AWG individually and collectively. It is at this point that the ungainly idea of the "Anthroposcene" becomes relevant to what now can be labeled the field of "Anthropocene Studies."

5.7 The Anthropocene and the Arts: Enter the "Anthroposcene"

As noted earlier, in 2016, the present author embarked on a research project to discover the extent and content of reporting on the Anthropocene in print media all over the world, accessed mainly online. One of the most surprising findings of the Anthropocene Media Project was the volume and scope of reporting Anthropocene-related arts events (see Sklair 2021, Chap. 13). While the largest numbers of these items appear in North American and Western European media, a significant volume of Anthropocene-related creative arts events appeared in media worldwide. Two articles published in the mid-2010s, by which time the Anthropocene concept had begun to penetrate the social sciences and the humanities, clarified some questions about the different ways in which the media, Earth System scientists, social scientists, and the rapidly growing community of environmental humanities scholars were dealing with the Anthropocene. Two journal articles introduced the concept of the "Anthropo(s)cene" (Castree 2015) and the "Anthroposcene" (Lorimer 2017), both explicitly drawing attention to the implications of the "geological" Anthropocene for the social sciences and humanities and vice versa. It is clear that most, if not all, members of the AWG were more or less comfortable with this development, though probably not with the new terms, however, spelled. However, other geological and Earth Sciences community members were uncomfortable with this apparent intrusion into their "hard sciences."

A rapidly growing literature on how creative artists engaged enthusiastically with various interpretations of the Anthropocene was forcefully brought home to the current author from an event in Brighton supported by Arts Council England in 2019 in which a group of local artists invited the public to "talk about the Anthropocene" and visit a multimedia exhibition in which the Anthropocene was interpreted aesthetically in many different ways. This provided on-site evidence that when artists in any medium take the Anthropocene as their subject matter, they must confront at least two enormous problems first, what Timothy Morton calls the "hyperobject" that the Anthropocene is (Morton 2013), and second, what we (human beings) can do about it. It is unclear if the Anthropocene or global warming (metonym) is the hyperobject. See Boulton (2016) for what might be called a "hypercritique" of Morton. Again, the confusion is not the point, Morton's widely read work draws attention to the Anthropocene.

References to the Anthropocene in media coverage of the arts display a variety of attitudes. It is usually explained in stock phrases like 'Age of Humans' and 'man's impact on the planet', but more overtly, dystopian sentiments are implied, as will be seen. A good guide to the historical and performative complexities of trying to express the Anthropocene artistically is provided by Robin and Muir (2015), who introduced the challenging idea of 'Slamming the Anthropocene', illustrated graphically with images of environmental hazards and reviews of events.

The following section summarizes some of the most prominent examples of media coverage of Anthropocene-related arts events and then reviews a selection of other items, organized by genre.

5.7.1 Anthropocene Outreach via Museums and Arts

As noted earlier, two major multimedia exhibitions in Germany, "The Anthropocene Project" at the Berlin Museum Haus der Kulturen der Welt (2022), and "Welcome to the Anthropocene: The Earth in our hands" in the *Deutsches Museum* (Möllers et al. 2015) in Munich illustrate how the AWG embraced the arts and how the arts embraced the Anthropocene. Damianos (2021) reports that HKW worked closely with the Anthropocene Working Group by providing a meeting place for its deliberations and funding for various Outreach activities, including an impressive multimedia website (see further). An early item on HKW appears in *Edición Impresa* (Argentina) in 2013 "Nature claims for justice," by the German journalist/poet Timo Berger, who interviews Paulo Tavares, urbanist and architect born in Brazil, teaching at Goldsmiths College in London. Tavares (2014) indicates the interdisciplinary nature of his work, testifying to the interest of HKW in crossing science-arts-humanities borders (see also Anderson 2015). Berger tells us that Tavares was one of the experts invited to HKW in Berlin for an international meeting to debate the future of our planet.

Business Report, a South African online news site, published an article in 2014 by the Washington-based environmental journalist Seth Borenstein, "Anthropocene, the age of humans: We're changing the Earth. There is no question about that, I've seen it from space" (a version of this article also appears in *China Post* in 2016). Borenstein reports that the American Association for the Advancement of Science is hosting an art exhibition, "Fossils of the Anthropocene," and that the AWG "ramps up its efforts to change the era's name with a meeting at a Berlin museum" (HKW). In 2015, the *Conversation* prints "The Anthropocene has been shaped by the media and our digital lives" by the communications theorist Jussi Parikka (2015). He writes: "the Anthropocene has given new inspiration to contemporary artists," citing the exhibition "Rare Earth" at the Thyssen-Bornemisza Art Contemporary in Vienna, and in Berlin, "HKW recently finished a massive multi-year project on the Anthropocene that curated exhibitions, talks, publications and a public visibility for a concept that has unified the humanities, earth and climate science." The *Egyptian Gazette* (nd) provides links to Anthropocene Curriculum—HKW, (inviting public participation) and to the ongoing multimedia "Anthropocene project" by Edward Burtynsky and his colleagues (see further). The German tabloid *Bild Zeitung* covers the HKW exhibition several times, for example, with a preview in 2012 reporting on the upcoming 2-year project by artists and scientists "to reflect on human's forming of nature," and again in "Experts want to introduce new age of the earth (*Anthropozän*)" in 2016. In Singapore, the *Straits Times* (2014) picks up the story from Reuters in Berlin: "Scientists prepare for a change of epoch to reflect humankind's deep impact on the planet." This article tells us that the AWG, being in no hurry to make important decisions, will report its conclusions in August 2016 to the geology congress in Cape Town. "Berlin's HKW, where the talks took place, is holding an exhibition of related artworks, videos and seminars." We are fortunate to have several commentaries on the exhibition by some involved scholars (Swanson et al. 2015, pp. 152–154, discusses this and other academic events).

An extensive digital archive exists, searching for "Anthropocene Project HKW" on the internet in 2020 brought up more than 400,000 results. As noted earlier, the Munich exhibition announced itself with a more direct title, "Willkommen in Anthropozän" illustrated on the cover of its book on the exhibition (see Fig. 13.1 in Sklair 2021).

El Clarín in "Technology is our second skin" (2014) profiles Helmuth Trischler, historian of science and technology and research director at the Deutsches Museum and director of the prestigious Rachel Carson Centre in Munich, who discusses androids and the origins of the Anthropocene in the agricultural revolution thousands of years ago. In the *Frankfurter Allgemeine Zeitung* (2014), it is reported: "We have arrived at a new geological age … traces of man are everywhere" and in the *Sydney Morning Herald* "Lightning rod for science" (2015). Both of these reports announce the Anthropocene exhibition in the DM. Also, in 2015, the *Newcastle Herald* (Australia) reported on *Deutsches Museum* (dubbed the world's largest museum of science and technology) and commented: "Having been around for so long, there's a certain tension between its historic exhibitions and those depicting the current state of science. On one hand, vast halls are populated by maritime and airborne craft of the past; on the other, temporary exhibitions such as 'Welcome to the Anthropocene'." *Abendzeitung* (2013) reports that people have been shaping the environment for thousands of years through agriculture, and that the special exhibition is dedicated to "Anthropocene, the Age of Man." The article tells us that the exhibition gives a sense of how much we humans have changed and adapted our planet in recent millennia and how powerful the influence of technical innovations has become. Visitors are invited to create visions of the future on a field of "1000 paper flowers." Critics have argued that these "flowers" neutralize the potentially dangerous impacts of the Anthropocene and contribute to the illusion of a "good" Anthropocene (see Bennet et al. 2016 on the case for a good Anthropocene, and for a critique, Sklair 2021, pp. 27-28 and *passim*). *Süddeutsche Zeitung* profiles the exhibition "Welcome to the Anthropocene" and explains what kind of technology is shown. "We live in man's vulnerable age. The Great Exhibition Anthropocene in Munich shows how mankind changes the globe far down to Earth's crust" (2015). Also, in 2015, the US-based magazine *Slate* (which covered the Anthropocene regularly) published "Bringing Postnatural History into View," a scholarly article on the domestication of species of plants and animals that mentions DM's "Welcome to the Anthropocene" in passing. More informatively, the New Zealand Geographic (2016) celebrates a long-lost recording of the call of the now extinct huia bird, in "part of Munich's Anthropocene exhibition helps us appreciate loss in the Anthropocene. But man also changes himself." In the *Guardian*, Robert Macfarlane cites this poignant story of the huia in his comprehensive article "The Anthropocene

Generation" (2016). He writes: "The DM in Munich is currently hosting An Anthropocene Wunderkammer," which it calls "the first major exhibition in the world" to take the Anthropocene as its theme. Macfarlane argues, "Anthropocene art is, unsurprisingly, obsessed with loss and disappearance." *Welt Am Sonntag* is one of the few media sources that mention Berlin and Munich together and put them into a broader context. In a thoughtful article in 2014 "Forget New Year's Eve! A new era has begun," Andreas Rosenfelder, head of the paper's Arts section, ranges far and wide to give meaning to the Anthropocene, "the hottest scientific debate these days," referencing the major exhibitions at HKW and DM. Scholarly opinion on these two exhibitions is divided. The conclusion reached by two historians of technology, reviewing the DM exhibition for the leading journal in their field, is highly critical: "the Anthropocene idea has prompted many scholars and activists to point out the radical environmental injustice of the epoch and to critique the capitalism that has led to it. We see neither in this exhibit. Instead, it offers a relatively benign vision of a changing planet. The change is not pictured as threatening, despite being rapid. The exhibit says Welcome to the Anthropocene, not Goodbye to the World You Knew" (Jørgensen and Jørgensen 2016). However, other scholars (Keogh and Möllers 2014; Robin et al. 2014) present much more positive evaluations of the Berlin and Munich events and, in general, the challenges that museums face to represent the Anthropocene aesthetically and symbolically.

Of the many other museums that hosted Anthropocene exhibitions, two others also receive significant coverage in the media. Calatrava's "Museum of Tomorrow" whose opening exhibition had strong Anthropocene themes, was labeled in 2016 by the Mexican paper *El Reforma* as "the new iconic building of Rio de Janeiro," also reported in *Extra El Clarin, Revista, Epoca, Le Soir,* and the *Hong Kong Economic Times*. The Smithsonian in Washington made the news when it officially recognized the Anthropocene. The museum official John Kress is quoted by the *Scranton Times-Tribune* (from an AP release reprinted in several other papers in the USA): "the Smithsonian is embracing the term because … for us it kind of combines the scientific and the cultural in one word," music to the ears of the AWG, but not always to the geological mainstream. The Smithsonian also appears in press stories similar to those in the media, such as *Los Andes, Borneo Bulletin*, and the *China Post.*

5.7.2 The Anthropocene Project: Burtynsky, Baichwal, de Pencier

Perhaps, the longest-running arts-related Anthropocene project is the mission of the Canadian photographer Edward Burtynsky to photograph human impacts on the planet, notably images of land, mining, energy, water, and industrialization. The *Houston Chronicle* (2016) reports that "FotoFest 16" will tackle the world's toughest environmental issues with its 16th biennial, featuring 34 leading photographers and mixed-media artists from nine countries. Burtynsky is highlighted but we do not learn much about his work, it is squeezed into a busy program. The *Guardian* (2016) with "Edward Burtynsky: the photographer finding art in rivers of toxic waste" tells us a great deal more, for example, that his "subversive activism ... is now more important than the artistry" and that a new film [the third in a series] called "THE ANTHROPOCENE: The Human Epoch," which Burtynsky made with the Canadian filmmakers Jennifer Baichwal and Nicholas de Pencier will investigate a current scientific theory, arguing that we are now living in the Anthropocene epoch, because "mankind's activities are shaping – and destroying – the Earth's ecosystems." *Maclean's* magazine in Canada in "The unjudging eye of Ed Burtynsky's camera" (2016) by Brian Bethune declares that "One of Canada's finest photographers considers his craft." This is a glowing assessment of Burtynsky's work, illuminated by spectacular photographs. Burtynsky is quoted: "How long can we be the top predator acting without concern for all the rest?" Notably, for the first time Burtynsky was photographing animals, illustrated by an iconic pyre of elephant tusks. The *Guardian* (2017) "In this new human age, the Anthropocene" quotes Burtynsky again: "We've reached peak everything" and follows up in 2018 with "The Human Signature: Edward Burtynsky's Anthropocene – in pictures." There are more spectacular images online (see Lauren 2018, *British Journal of Photography*, and for an anthropological perspective on the exhibition, see Evans and Ryan 2019).

There was more coverage of the Anthropocene film after its release in Canada in 2018, for example, in the *Ottowa Star* (which had previously given space to climate change deniers). Articles documenting the evolution of the project also appear in the *Toronto Star* and the *Ottawa Citizen*. In the *Citizen,* Burtynsky explains that Jennifer [Baichwal] said, "Why don't we take this word, Anthropocene, and build a film and book and everything around that? Burtynsky was sceptical... Every time I spoke to audiences I would say, has anybody heard this word? and maybe four hands would go up. Now it's on the side of buses in Toronto." In the *Toronto Star,* we find: "Sprawling Anthropocene project shows humanity's enormous impact on the planet." The art critic Murray Whyte says: "Burtynsky's pictures have always held a terrible beauty. His compositions veer close to the abstract in their capturing of horrendous damage ... They're gorgeous first, horrendous later, and that's surely the point'. De Pencier is quoted: "Someone called us the three horsemen of the apocalypse ... I really hope that's not the case. But we can't claim neutrality anymore. We used to say this is not a polemic, and you can draw your own conclusions." Whyte (who is not entirely uncritical of some aspects of the exhibition) continues: "For Burtynsky, who has worked along thematic lines tied to specific industries — railway lines carved into the landscape; mines; farming; urban sprawl — *Anthropocene* presents as an unintended catch-all term for his entire career." This is a thought-provoking conclusion that might also fit some environmental humanities scholars. And in the *Daily Mail* (2019), "Tire towers, mesmerizing mines and deforestation enthuses on Burtynsky's amazing aerial photos showing how humans reshaped the planet." The Anthropocene film exhibition in the National Gallery of Art in Ottawa and the Art Gallery of Ontario in 2018 gave visitors an opportunity to express, in a word on a digital board, how the exhibition made them feel: 23% felt worried, 17% angry, 9% unconcerned or suspicious. It is difficult to keep track of the many voyages of this multimedia project. The Anthropocene film premiered at the prestigious Toronto International Film Festival and has been screened at many theaters and galleries across Canada, the USA, France, Italy, Glasgow, and elsewhere. The project (film, photographs, and installations) has traveled around North America, Flowers Gallery in London, and to Bologna in 2019. A complementary coffee table book was published by Steidl (enhanced digitally), with commentary from two AWG scientists (Zalasiewicz and Waters), new work from Margaret Atwood, and another volume served as a catalog for the exhibition.

5.7.3 La Repubblica: Anthropocene and the Arts in the Italian Media

A distinctive feature of coverage in the Italian media (over 100 items found) was the prominence of Anthropocene-related arts events in the center-left daily *La Repubblica* (27 of 46 items found). These ranged from "One hundred films from all over the world" (2005) to a report on the Ecovision film festival in Rome. The Anthropocene reference is for Nobel prize-winner Paul Crutzen, famous for his studies on the ozone layer, presenting his new book. The title of the event is "Welcome to the Anthropocene." In 2013, the first in a series of Anthropocene-related musical events is announced, "Leu Music in Casalborgone" reporting, rather obscurely, "hypothetical finding of a black box dispersed in the Anthropocene that brings to light extinct sounds, images of lost places, voices of daily life and excerpts of conversations." Several articles announce events involving the multimedia musician Marco Paolini (one of whose concerts, "The Anthropocene" at the Roma Europa festival in 2017, (available on You Tube), and on a music festival in Barcelona, "Dance and sounds of the Earth: at the Sonar you dance the era of Anthropocene" (2016). The reporter wryly observes: "Welcome to the era of Anthropocene. The sentence was not written on any door or access gate here at the Sonar in Barcelona, yet"; an Anthropocene recital by the eminent Russian pianist, Arkadij Volodos, at the Carlo Felice theater in Genova is announced (2016); "Giraldi between man and nature" (2017) is a sympathetic review of the exhibition "cosmetics of the Anthropocene" featuring Piero Giraldi, one of Italy's leading politically engaged artists, his "nature rugs" attracting great critical interest. Anthropocene-related museum events also feature in the coverage in *La Repubblica,* notably a wide-ranging review of a major exhibition "Maxxi in Rome, art and science of the cosmos" in 2017. The catalog explains: "From gravitational wave detectors to an installation on the world after the Anthropocene: on display are instruments and works by scientists and artists. Closer than you think." A subsequent event in 2018: "Gravity. Imagine the universe after Einstein" is previewed, **"the** *Cosmic Concert"* **by Tomàs Saraceno** features an enormous image of a spider that moves on the canvas to reflect the texture of space. **The curator observes that art and science are closer realities than you think.** Saraceno, for example, was the first artist with access to the CERN laboratories, and his work in the Maxxi museum on *Aeroke* (reflecting spheres suspended in the air, surrounding the Cassini antenna) represents how the artist imagines the world after the Anthropocene. The final selection from the Anthropocene arts archive in *La Repubblica* reports the last stage of an exhibition at the science museum in Trento. "The ecosystem and its life cycles" (2016) is dedicated to the Anthropocene, apparently labeled "Extinctions. Stories of catastrophes and other opportunities." The following section is a small selection from a large archive, organized by genre, demonstrating the global impact of Anthropocene Outreach.

5.7.4 Genres

Art Exhibitions The third "End of the World Biennale" in the city of Ushuaia, in southern Argentina was reported in *El Universal* (Mexico) and in the Argentine papers *Ambito Financiero* and *La Prensa* in 2011. Works by 100 artists from all five continents were exhibited, inspired by the idea of the Anthropocene. Despite the ambiguous title (Ushuaia markets itself to tourists as the city at the end of the world), the consensus is that the exhibition is a great opportunity to demonstrate that another world is possible if compromise and common sense prevail. The article concludes that the concept Anthropocene is not just gaining attention in the scientific spheres but also within artistic circuits. The growing number of art exhibitions related with the topic might be showing the beginning of a social process, where arts help to pose new questions about our common future. *Hangzhou Daily* "Taipei Biennale, there is a big world in the small play" (2014), reports that the theme is "drama and acceleration," and the proposition is "Anthropocene." The works of 52 artists from around the world are on display here. The Luxembourg tabloid *Le Quotidien,* "Casino: activities continue during building works" (2016) tells us that 27 young artists from Strasbourg exhibit their works on the theme of the anthropocène, or rather on its affects as a principal actor in climate change. *Daily Star* (Bangladesh, 2016) "Fragments of the Anthropocene: Bengal Art Lounge" describes the works of two artists from Dhaka, from their exhibition about the loss, which the onset of the Anthropocene has caused. In 2012, the *Boston Globe* prints several items on the artist known as Swoon, in front of her installation "Anthropocene Extinction" at the prestigious Boston Institute of Contemporary Art. The artist is pictured speaking at the museum's Teen Night with some Boston high school students. In Argentina, *Página 12* "Nature: refuge and resource for humans" (2017) briefly summarizes an art exhibition including "the origin of an idea, the Anthropocene" along with utopias-dystopias, sustainability, ecological engagement, activism, ecosystems, new worlds, and new natures. The item concludes: "More evidence on the growing number of art exhibitions and events placing the Anthropocene as a central axis of curatorial projects." Finally, in Scotland, the *Greenock Telegraph*, reports "Grove Park hosting green residency this weekend" (2016) where eight Scotland-based artists will spend a weekend looking at the connections between the arts and a more environmentally sustainable society. The event is the third annual artist residency organized by Creative Carbon Scotland. CCS director Ben Twist is quoted: "The Anthropocene marks a significant shift in how we understand humanity's role and responsibility in relation to issues such as climate change ... Exploring this topic during the residency offers an important opportunity for Scotland's artists to consider how their practices could contribute to a wider cultural movement towards a low-carbon future." Also quoted is Lex Braak, director of the Van Eyck Institute in Maastricht:

> I am sure my understandings of the Anthropocene will change and deepen from the engagement. Residencies are wonderful settings in which to create shared understanding across disciplines and practice settings.

Photography In 2017, the *Washington Post* reported "Climate change exhibit to open near Trump's house," profiling the work of photographer Justin Brice Guariglia, who brings the catastrophic results of climate change to Trump's own doorstep, in Florida. His exhibition "Earth Works: Mapping the Anthropocene" was covered all over the world, for example, in *Egypt Today* and Taiwan's *United Daily Times*, featuring images of melting glaciers that Guariglia collected while on a NASA mission to Greenland. The *New York Times* profiles his work in "A Man on an Eco-Mission in Mixed Media" (2017). He tells the *Art Newspaper*: "One of the unique things an artist can do is to help shape realities and, at the same time, have a political and social message. Artists today really have to use art to carry an important message forward." An essay by the curator (Wride 2019) explains the unique qualities of Guariglia's work. Like many artists, he has been inspired by Greta Thunberg, reported in an article on *Artnet News* (Rea 2019).

The Canadian magazine *Maclean's*, "Postcards from the edge of the earth" (2016) reviews a photography exhibition, chronicling the environmental movement from the 1960s to the present. The curator says that the title suggests falling off a cliff, perhaps one that mankind blasted itself. "Maybe human beings are not the strongest species on Earth. It's something to say, beware." *La Crosse Tribune* (Wisconsin), "UW-L photographer's Winona exhibit comments on a precarious environmental future" (2016), reviews "Anthropocene," pictures of mass-produced disposable goods such as candy and cheap trinkets. The artist explains: "the aesthetic conveys the turmoil of an inevitable ending that we are aware of on some level, and that breeds a very complex anxiety." *National Business Review* of *New Zealand* reports how a collection of photographs of the industrial past is transformed by Chris Corson-Scott into an exhibition, "Dreaming in the Anthropocene" (2017). On photography after the Anthropocene, see Kember (2017).

Music In the *Seattle Times* "Say goodbye to 2016 with a song" surveys the year in music with reference to Nick Cave's song "Anthrocene," which "felt appropriate because it wraps up the loss felt this year and the severe impact of humans on the environment." The writer explains that the word "Anthrocene" was first used instead of the now more common Anthropocene in 1992 by science writer Andrew Revkin in his book *Global Warming* to describe a "geological era significantly impacted by humans." In England, the *Oxford Mail*, "Death of Hi Fi creates a 'squall' of sound" (2012), profiles a local group and their new album, "Anthropocene." Lead vocalist Andy says: "There's a beauty in darkness in some ways and I get a lot of solace from it. I was probably the only person to read Cormac McCarthy's *The Road* and find it uplifting." The new album "Miss Anthropocene" by the enigmatic Canadian pop star Grimes is reviewed widely in the media in 2020. Wodak (2017) provides an acute analysis of climate change and Anthropocene narratives in popular music. In the Danish *Politiken* "Classical music in antropocæn time" (2016) connects the "Voice of the Whale" from 1971, a key work in environmentally conscious composition music with the Anthropocene. Finally, several media outlets in the USA report that the Norton Museum in Florida hosted more musical events inspired by the photography of Justin Brice Guariglia (see aforementioned).

Opera *DV* (Iceland, 2016) reports the contemporary opera "UR" by Icelandic composer Anna Þorvaldsdóttir based on the idea that Man considers himself above the environment, affecting the Earth in an unprecedented way. She is quoted: "Because of this, theories have emerged that have us living in the Anthropocene [*mannöld*], an epoch of human impact in Earth history. This can be seen most clearly in places where people have lived for thousands of years in a very close relation to the harsh natural forces, like on Greenland where the concept work for the opera was partially carried out. The glaciers are melting, the oceans are acidifying, and the livelihood of animals and human societies is crumbling." And the Scottish Opera company production "Anthropocene" featuring an unfrozen ice maiden was reviewed by several Scottish newspapers in 2018.

Dance Media reporting of Anthropocene-related dance is generally uninformative, usually listing title of performances but not their content. The Bangladesh *Daily Sun* (2016) reports on the Jahangirnagar University dance troupe and its 11-day-long festival to celebrate three decades of its journey. The Anthropocene connection is obscure. *Salzburger Nachrichten* (2016) advertises "Anthropozän," an urban dance performance based on the influence of man on the Earth, while *Süddeutsche Zeitung* (2017) "In the crosshairs of the gods" reviews a ballet that deals with the end of humans on earth and their self-destruction as a metaphor for the Anthropocene. In Australia, the *Canberra Times*, "Art gets in on the act of climate debate" (2009) reports on a climate change and sustainability forum and the performance of Mirramu Dance Company's "River," implicating the Anthropocene in the drying up of the Murray River. An optimistic activist is quoted: "It is fixable … Clear vision is more important than deep anger."

Sculpture *Grapevine*, the Icelandic anglophone weekly and online news site, profiles the sculptor Anna Rún Tryggvadóttir in "Everything is in Motion, nothing is Solid" (2017), a long and especially instructive article on artistic practice in the Anthropocene by the Berlin-based writer and performer, Steindór Grétar Jónsson. Tryggvadótti is also based in Berlin, her exhibition "Garður" (Garden) is at the Reykjavík Art Museum. The main components are Icelandic rocks, which the artist sculpts and treats. Suspended above these rocks, containers of colored liquid are rigged to drip every hour, watering them and changing their appearance. Slowly, they transform in color and texture as chemical compounds on the rocks, like salt, react to the stimuli. Her central philosophy is rooted in the state of absolute authority over a situation while being completely powerless to control it at the same time. The situation of the artist, she says, involves "how we position ourselves toward nature and how society constantly demands that we encroach upon it ... It's a twisted spot we're in. Scientists are calling this era the anthropocene—this period of time from when mankind started to significantly impact the world. We tend to define everything from the perspective of humans and how we understand and perceive things. I try to mirror our function down to a material or even molecular level, through the use of matter we consider as being inanimate." Her techniques inevitably lead to the artwork morphing from day one up until the final day of the exhibition. Reflecting on her continued connection with Iceland, she explains, "Even though Icelanders are raised in a very European society, we face wild and crazy weather and unstable elements, it's so different from living in a big city, a manufactured area for human existence, built to service our needs. We're at a point in human history where we have to balance this with the idea of conserving the resources needed to create this kind of reality. In Reykjavík you can look around and see nature, the mountains, the ocean, almost 360 degrees. In Berlin, I don't have any of this. Being here has stirred something in my subconscious." This article is quoted at such length because it is almost unique in showing how the work is manifestly influenced by and centered in the Anthropocene. *The State Journal-Register* (Springfield, Illinois, 2014) is one of many media to report on the art of the British sculptor, Jason de Caires Taylor: "The only way to see this incredible museum in Mexico is by scuba diving." The installation is of 500 underwater sculptures off the coast of Mexico's Riviera Maya, via the Museo Subacuático de Arte. One exhibit is called "Anthropocene" a sculpture of a Volkswagen car made to house lobsters. A picture of this sculpture adorns the cover of Heise et al. (2017).

Theater *Le Soir* (France, 2017) gives few details in "The Yes Men: spreading some joy at la Fabrique de Théâtre à Frameries, sponsoring a studio dedicated to the anthropocene." *Süddeutsche Zeitung* (2017) under the rather obscure title "This is the molecular genetics," profiles the German dramatist Konstantin Kuspert, who is known for his dystopian (and sometimes utopian) plays and controversial topics. Kuspert talks about the Anthropocene, saying: "it is outrageous how it is exploited by the media but there seems to be no change in people's mentality." *Canberra Times* (2016) reviews playwright Hannie Rayson's "Extinction," explaining "The sixth great dying – the so-called Anthropocene extinction – is the one we're living through now. It covers the loss of biodiversity since humans emerged as a super-predator, and although scientific consensus on key questions remains elusive, there's broad agreement our ecological impact on the planet is global and severe." But nothing, the reviewer concludes, "offers relief from the playwright's thin take on a problem thick with dramatic potential." This dramatic potential appears more realized in the *Calgary Herald* review of "Extremophiles" (2018), Georgina Beaty's forward-looking apocalyptic and absurd solo show, inspired (if that is the word) by resource extraction in Alberta, set in a desertified and abandoned Canadian North in the year 2020. She is quoted: "I am fascinated by the Anthropocene Age, the time in which we are living, where humans have permanently altered the earth's natural systems. Climate change is an unavoidable undertone in my work." Once again, we see climate change as a metonym for the Anthropocene, which is understandable in the context of Alberta's tar sands and wildfires, localizing the Anthropocene. Also from Canada, the *Edmonton Journal* "Swallow returns to Edmonton to ask hard questions about our future" (2016) gives us a lengthy report on the Frente (as in "out in front") Theatre Collective. Playwright Leslea Kroll's play, originally conceived as a comedy for a melting globe, chronicles the demise of a fictional island entirely swallowed up by the rising ocean. Kroll's sense of urgency intensifies on a day when the city is full of Fort McMurray evacuees and wildfire smoke. The Portuguese magazine *Sabado* in the curiously entitled "Your Majesties, Welcome to the Anthropocene" (2015), reviews a theatrical performance "Talk to the Demon" in which Wim Vandekeybus works with a team of six performers and a child. Vandekeybus explains: "We humans are desperately busy with giving sense to everything that happens between two mysteries - birth and death … We all create demons inside ourselves to fight out our inner struggles. Good can't exist without bad'. The reviewer says 'Talk to the Demon is a piece without music. In that sense it's very scary, very naked, but at the same time also pure and touching."

Poetry In the South African media platform, *Netwerke 24*, we find "T. T. Cloete's latest collection is a fascinating study of the relationship between God, human and nature" (2014) by Andries Visagie (Professor in Afrikaans at the University of Pretoria). This poetry review is interesting because it is an early reference to the Anthropocene in Afrikaans literature. The work attempts to reconcile science and religion, using nature as a lens through which to experience humanity's effect on the world. "seeking meaningful ways to live their faith in the world." The section on the Anthropocene outlines an oblivious image of modern humans who have to sacrifice quality of life because of its rapid growth and its insatiable urge for consumption. *Smoky Mountain News* (Waynesville, North Carolina) publishes "Book of poetry has a disturbing beauty" (2017), a thoughtful review of "Anthropocene Blues" by local author John Lane, "these poems are told from the perspective of a traveling geologist as he sings the blues of our epoch… about how we humans are taking ourselves down along with everything around us. This is not an apocalyptic book about the collapse of human existence on the planet, but it is a wake-up call, which has been the poet's role for many millennia as the harbinger of good and bad news." Local landscapes feature strongly in these poetic reflections, echoing the "localizing of the Anthropocene."

The year 2016 was a bumper year for Anthropocene poetry, beginning in *The Telegraph* (London) with a review of "Poetry and the Anthropocene: Ecology, biology and technology in contemporary British and Irish poetry" by Sam Solnick (an academic at Liverpool University): "This book asks what it means to write poetry in and about the Anthropocene, and argues that the Anthropocene requires poetry and the humanities to find new ways of thinking about unfamiliar spatial and temporal scales, about how we approach the metaphors and discourses of the sciences, and about the role of those processes and materials that confound humans' attempts to control or even conceptualise them." In "Poems help comprehend the incomprehensible" the popular Danish newspaper *Politiken* argues that "Poetry in the Antropocæne [sic] era deals with both dystopian and utopian dreams, with future disasters and inhumane scale conditions – in short, everything that science can describe, but does not make sense." *Süddeutsche Zeitung* brings the anthology *Poetry in the Anthropocene*, edited by Anja Bayer to the attention of its readers. The *Vancouver Observer* opens its pages in to the local academic and poet Stephen Collis, in "Poetry against the pipelines.' Collis eloquently explains the relevance of poetry to real life, "Ive written about the desire for change in a world that is already rapidly changing — in ways that alarm us, or at least should alarm us." He discusses the curious fact that his poetry was used as "evidence" in an Air India terrorism trial, and his book of poetry about walking in the Anthropocene, "along pipeline routes and proposed pipeline routes, and the strange and threatening paths they cut through our environments" notably the Alberta tar sands, described as the construction of a vast desert in the middle of the boreal forest. He speaks about the power of "poetic documentary … mostly I find poetry is best at digging into motivations, into the feelings driving us to act and resist." *Al-Iraqiya* tells us about an American poetry magazine's latest issue focused on climate change, mentioning Craig Santos Perez's poem "Halloween in the Anthropocene" (another example of the way in which climate change is used as a metonym for the Anthropocene) and another example of what we might call "The Anthropocenic Imagination."

Cli-Fi (Climate Fiction) An article in *Conversation Africa* "Extinction or survival: how storytellers explore the ethics of colonising other planets" (2017) by Siobhan Lyons (2017) of Macquarie University in Australia, provides a splendid introduction to Anthropocene-related cli-fi. She begins: "Interplanetary colonization was once the stuff of science fiction but now there are plans to colonize Mars. How have film makers and writers dealt with our rapacious Anthropocene age?" Lyons cites Kim Stanley Robinson's influential *Mars Trilogy*, Andy Weir's novel (and subsequent film) "The Martian," and the film "Interstellar" (see further), showing how these fictions raise questions about human survival, making an interesting comparison between the 1951 version of the film "The Day the Earth Stood Still" and the 2008 version, the earlier focused on the nuclear threat, the later remake in 2008 on environmental catastrophe. Lyons argues that the book and film of Shute's *On the Beach* and Ballard's *The Drowned World* (and many others) pose questions about whether humanity deserves to survive, remarking on Woody Allen's "Ozymandias Melancholia … the artist's job is not to succumb to despair, but to find an antidote for the emptiness of existence" (some may consider that this sounds apt for the Anthropocene.) Lyons concludes: "But when the prospect of human survival intrudes upon the natural environment of other planets, which would be best left alone, the idea of colonizing other planets becomes unethical" (2017). There is a flourishing Anthropocene-related scholarly literature on cli-fi, notably, Johns-Putra (2010, 2016), and Milner (2018: part 3).

Seattle Post-Intelligencer (2016) in a review of *Barkskins* by Annie Proulx proclaims: "And if we're lucky enough to survive the Anthropocene we've seemingly wrought, then *Barkskins* will surely survive as the crowning achievement of Proulx's distinguished career, but also as perhaps the greatest environmental novel ever written." In *Times of Israel* "Literary Critics Probing the Rise of Cli-Fi Novels in the 21st Century. Cli-Fi, Sci-Fi, We All Cry, The End is Nigh: What Cli-Fi Novels Say About the Anthropocene" (2017), the writer suggests that Adam Trexler (2015) led the way with his book *Anthropocene Fictions*, a sentiment most scholars share.

Film There are two distinct types of Anthropocene films. The first comprises documentary films made explicitly to explain the Anthropocene to various publics, for example, the short but much referenced "Welcome to the Anthropocene" for Rio +20, the longer "Welcome to the Anthropocene" by Steve Bradshaw, the Australian Broadcasting Corporation documentary "The Anthropocene: The age of mankind." The best-known example of this genre is probably Louie Psihoyos' award-winning Discovery Channel documentary "Racing Extinction" in the context of the Anthropocene, reviewed in the Argentine *La Nación,* the Bengali magazine *Outlook, La Tercera* in Chile, Pakistan's *News International*, Portugal's *Grande Reportage, Seattle Times*, and other media. The detailed evaluation in the *Cornell Chronicle* (the university paper) concludes that the film "offers a series of recommendations about what we as individuals can do to try to slow down the tidal wave of extinction, whether it's skipping meat for a day or switching to a greener source of energy. Science is important but it's not the answer. The answer lies in human behaviour." However, not everyone liked it, Naomi Klein writes a penetrating critique of "Racing Extinction" in *This Changes Everything* accusing it as "eco-opportunist" (see Truscello 2018).

While some feature films make little or no explicit reference to the Anthropocene, but their reception has been framed as "Anthropocene" events. Salmose (2018) labels this "Hollywood's apocalyptic sublime," and Fay (2018) adds historical depth—both referencing the Anthropocene.

The impact of Roland Emmerich's legendary "Day after Tomorrow" has been analyzed in detail, the catastrophe is a new ice age brought about by global warming (Reusswig and Leiserowitz 2005; Hart and Leiserowitz 2009). For those skeptical of the place of film, a review of the film by the distinguished climatologist Stefan Rahmstorf (concludes: "I think it would be a mistake and not do the film justice if scientists simply dismiss it as nonsense" (Rahmstorf 2004). This is an important statement in terms of planetary consciousness as well as popular culture in the Anthropocene not just for "Day after Tomorrow" but for others like *Interstellar*. This argument is expanded by Michael Svoboda (2014).

Interstellar (2014) raises issues of an increasingly uninhabitable planet and the search for a more liveable one (see Dickens 2015; Powell 2017; Andersen and Nielsen 2018). If it ever gets made, the film of the legendary Kim Stanley Robinson's

"Mars Trilogy" will surely join this list. And Johnson (2014) provides an exhilarating introduction to Manga in the Anthropocene.

Multimedia The *Lebanon Daily Star* "The aesthetics of extinction" (2016) reports an exhibition "Let's Talk about the Weather" at the Sursock Museum in Beirut, including works by 17 artists from Lebanon and beyond, supplemented by a public programme of talks, workshops, films, and guided tours of toxic sites in Beirut, plus "Elements for the World," a series of five English and Arabic-language publications on Stone, Water, Wood, Fire, and Sky edited by Ashkan Sepahvand. Noting that most of the work linking the Anthropocene to contemporary art comes from the north, the curators, Nataša Petrešin-Bachelez and Nora Razian, see this show as speaking from the global south. "The touchstone for these labours is the Anthropocene." The daily *Seattle Times* (a socially conservative newspaper) has six Anthropocene-related articles, five of which are cultural reviews, with little substantive discussion of the Anthropocene. An exception is "Bloom and doom: Pokémon Go-like art project explores climate change" (2016) reviewing "Gardens of the Anthropocene" by Tamiko Thiel, who, with help from University of Washington scientists, creates augmented-reality art, linking to articles on "The glaciers are melting. Sea levels are rising" and "Seattle's summer of 2016 is breaking heat-wave records." The journalist remarks that this "makes it a grimly perfect time" to visit the exhibition, also listed in the *Seattle Post-Intelligencer*. The Danish *Kristeligt Dagblad* prints a very long article on a very broad event, "Nature and the Antropocæne man" (2017), part of the multimedia ARoS museum Triennial, a 4 km long art zone along the Gulf of Aarhus illustrating the human relationship to nature. *Star Tribune* of Minnesota prints what is probably the weirdest title we found, "What if you could understand climate models better by eating them?" (2016), on the work of Jonathon Keats, experimental philosopher and artist, who proposes "squishing all the continents together" into "anthropocenic sorbets." He calls the process "data gastronification." *Extra el Clarin* (2017) adds the ideas of Dr. Joachim Krausse (Humboldt University, Berlin) on anthropocene cuisine. The Scottish magazine *Skinny* reports the collaboration between Universidad Nacional de San Argentina and C-EENRG at Cambridge University, a multimedia experience of Earth's rotation in real time showing "the new geological epoch that we call Anthropocene fleshing out the unprecedented humankind footprint on our planet" (part of the Edinburgh Fringe festival). The *Skinny* also reports the Anthropocene-themed exhibition "Grike" at the Edinburgh Fringe. In an unusual case of contrarianism, the *Times-Picayune* (New Orleans) reports "Ecozoic exhibition explores our connection to nature through plants, seeds, soil" (2016) explaining that the curator of the exhibition, Bobbe Besold, makes a case for the term Ecozoic Era as opposed to Anthropocene epoch. The *Houston Chronicle*, "These bizarre and beautiful Persian rug-style images are made from Google maps" (2013) explains how the artist David Thomas Smith is putting on an exhibition called "Anthropocene." *Washington City Paper* reports that local artist Rachel Schmidt is calling her exhibition "Daydreams in the Anthropocene." The government-owned paper *Rossiyskaya*

Gazeta reports, "Intermuseum-2019 festival opens in Manege" (2019), a major theater, multi-site, multi-genre celebration, including "Welcome to the Anthropocene!" in the Perm Museum of Local Lore. *The Herald* (2015) "10 things to do in Dublin this weekend" picks the "Riddle of the Burial Grounds," a free, international exhibition motivated by one of our planet's major problems—the markings and warnings around nuclear burial sites. The journalist comments: "Did you know that mankind is entering the 'Anthropocene' era? Nope, me neither. I guess we'll just have to allow the visual arts to educate us further, and Riddle of the Burial Grounds has it covered via the mediums of sculpture, film, photography, documentary, science-fiction, and imagined futures." And finally, *sairblog.files* (2017) gives us the impressions of the director of a Japanese art and science interdisciplinarity initiative, who reports that what appeared most different upon arriving in Europe was the difficulty of finding an art exhibit or academic event where the "Anthropocene" (人新世) did not figure, a fitting conclusion to this cornucopia of artists of all sorts striving to express the Anthropocene aesthetically.

Japanese media provide a small number of Anthropocene-related arts events, for example, from *Japan Times*. "Darren Almond: All things pass" (2014) reports a video exhibition, part of which was labeled "Anthropocene," with short films "Sometimes Still" and "All Things Pass" described as "precisely choreographed multi channel installations at an architectural scale. They are at once figurative and abstract"; "Curator Okwui Enwezor tackles grim realities at Venice Biennale" (2015) by Cameron Allan Mckean, formerly cultural editor of the *Japan Times*, latterly a PhD student in Anthropology at Deakin University in Australia, begins: "Ugly, joyless, aggressive, didactic, morose, self-righteous, unpleasant; these are just some of the words used in the press to describe the recently opened 56th Venice Biennale in Italy." This is connected with "an attempt at articulating the horror and sense of dislocation we have with conceptualizing the scale of our new geological age, which many are calling the Anthropocene"; and a *Newsweek Japan* article by Ozaki Tetsuya, editor of a cultural magazine, discusses art and aesthetics in the Anthropocene in 2016. In the same year, "Black illumination: the unhuman world of Junji Ito" describes in graphic detail (no pun intended) the artist's manga "Long Dream" (now a TV film), referencing H.P. Lovecraft (who also influenced Donna Haraway), other masters of Japanese horror manga, even hints of Hieronymus Bosch or the much older tradition of the Buddhist jigoku-zoshi (hell scrolls). Tetsuya connects this with the anthropocene as our human-influenced epoch has been labelled, arguing that "we seem to be embedded in the world and estranged from it at the same time. And the most difficult thought is that the world, the cosmos, even our own bodies, all be indifferent to our daily wants and desires. It is this indifference, this stark and shadowy blankness, that is both alluring and terrifying. And this is the terrain of Ito's manga, where, for a brief moment, we glimpse something on the order of deep time and the scale of the unhuman." This is one of the few articles that successfully connects the art form with a meaningful Anthropocene imaginary.

5.8 Conclusion

It would be understandable if readers were not convinced that all the examples of "Outreach" identified in this chapter were entirely the work of the AWG or were actually cases of "Outreach." Readers might also be irritated by the long list of selected newspapers worldwide with Anthropocene-related material. Similarly, there is a long list of disciplines in which we find Anthropocene-related content, countries, and languages. The detailed descriptions of genres of Anthropocene-related creative arts might not seem very relevant to the activities of the AWG. My justification for concluding the chapter in this fashion is precisely to draw attention to how the geological Anthropocene community, especially most of the members of the AWG, opened their doors to all and sundry and, just as importantly, how so many people outside the geological community rose to the challenge of the Anthropocene notion and made an effort to communicate what it meant to them in terms of their skills and points of view. In this process, the place of "Outreach" is fundamental.

The conception of "Outreach" here is sociological, having more in common with its meaning for business, mostly persuading potential customers to buy products, not exactly "advertising" or "hidden persuaders," but not entirely different from these phenomena. Can we consider the Anthropocene notion as a peculiar product that is too complicated and arcane to be on shopping lists? Some geologists, skeptical of the Anthropocene notion, might agree that the analogy is apt. Two geologists published a paper in 2012 that seemed to resonate within the geological community. It suggested that the Anthropocene might have more in common with "pop culture" than stratigraphy (see Autin and Holbrook 2012). This publication was followed by Finney and Edwards (2016) in *GSA Today*, with a somewhat more ideological tint, entitled "The Anthropocene' epoch: Scientific decision or political statement?" Finney and Edwards draw attention to an editorial in *Nature* in 2011: "Geologists are used to dealing with heavy subjects, so who better to decide on one of the more profound debates of the time: does human impact on the planet deserve to be officially recognized? Are we living in a new geological epoch—the Anthropocene?" The editorial answered the questions as follows: *Official recognition of the concept would invite cross-disciplinary science. And it would encourage a mindset that will be important not only to fully understand the transformation now occurring but to take action to control it. ... Humans may yet ensure that these early years of the Anthropocene are a geological glitch and not just a prelude to a far more severe disruption. But the first step is to recognize, as the term Anthropocene invites us to do, that we are in the driver's seat.* (Nature - Editorial 2011, p. 254). Responding to the editorial Finney and Edwards write: "The term Anthropocene is widely used. In its latest iteration, it refers to the present, when human impact on Earth's surface, atmosphere, and hydrosphere has been deemed to be global. International organizations; national, regional, and local governments; non-governmental organizations; and industries have taken steps to mitigate and remediate the impact where its nature is judged to be deleterious. Nevertheless, human impact is immense and potentially increasing. But, the question is: Should the Anthropocene be approved by the ICS and ratified by the IUGS as an official unit of the ICS International Chronostratigraphic Chart?" The remainder of this article addresses only the technical geological issues, implicitly arguing that the essential questions brought forward by the Anthropocene notion are technical and bureaucratic, not political. There is no suggestion that the Anthropocene might represent any existential risk to humanity. It would be erroneous to suggest that the AWG advocated a position of existential risk or that its members paid less attention to the science than Finney and his colleagues in the numerous geological committees deciding on nomenclature. However, my argument is that AWG's outreach, especially its commitment to interdisciplinarity, kept the door open to the politics and the risks of the Anthropocene. AWG's outreach via the media, as illustrated by the number and volume of newspapers publishing material on the Anthropocene, by the openness to scholars and creative artists of all persuasions, by explicitly encouraging engagement with lay people, played a crucial role in kick-starting the belief that the Anthropocene did represent something that humanity should not be welcoming.

In the 2014–2015 AWG Newsletter, Jan Zalasiewicz, in his Chairman's Remarks, makes this statement: "We must, therefore, determine, what is to become of us post-2016. Should we disband, or should we stay together as a body (officially or unofficially) to continue to pursue our study of the Anthropocene (perhaps within a different framework, and with a different Chair)? Given that we have developed something of a positive and productive spirit, my feeling is that it would be good to carry on this work in some way. However, this is very much a collective enterprise, with collective decision-making. So ideas welcome! Jan Zalasiewicz." In 2019, Zalasiewicz resigned as chair of the AWG, to become Chair of the Subcommission on Quaternary Stratigraphy. He would be entitled to believe that even if the intradisciplinary nomenclature issue had been stalled or even aborted, at least with his leadership, the term Anthropocene had, indisputably, been established in all the major languages and across almost all scientific disciplines and beyond.

References

3rd Nobel Laureate Symposium on Global Sustainabilty. 2011. The Stockholm memorandum: Tipping the scales towards sustainability. *Ambio* 40 (7): 781–785. https://doi.org/10.1007/s13280-011-0187-8.

Akin, Heather, and Ashley R. Landrum. 2017. A recap: Heuristics, biases, values, and other challenges to communicating science. In *The Oxford handbook of the science of science communication*, ed. Kathleen Hall Jamieson, Dan M. Kahan, and Dietram A. Scheufele, 455–460. Oxford: Oxford University Press. https://doi.org/10.1093/oxfordhb/9780190497620.013.48.

Andersen, Gregers, and Esben Bjerggaard Nielsen. 2018. Biopolitics in the Anthropocene: On the invention of future biopolitics in Snowpiercer, Elysium, and interstellar. *The Journal of Popular Culture* 51:615–634. https://doi.org/10.1111/jpcu.12689.

Anderson, Kayla N. 2015. Ethics, ecology, and the future: Art and design face the Anthropocene. *Leonardo* 48:338–347. https://doi.org/10.1162/leon_a_01087.

Autin, Whitney J., and John Holbrook. 2012. Is the Anthropocene an issue of stratigraphy or pop culture? *GSA Today* 22:60–61. https://doi.org/10.1130/g153gw.1.

Benner, Susanne, Gregor Lax, Paul J. Crutzen, Ulrich Pöschl, Jos Lelieveld, and Hans Günter Brauch. 2021. *Paul J. Crutzen and the Anthropocene: A New epoch in Earth's History*. Cham: Springer. https://doi.org/10.1007/978-3-030-82202-6.

Bennet, Elena M., Martin Solan, Reinette Biggs, Timon McPhearson, Albert V. Norström, Per Olsson, Laura Pereira, Garry D. Peterson, Ciara Raudsepp-Hearne, Frank Biermann, Stephen R. Carpenter, Erle C. Ellis, Tanja Hichert, Victor Galaz, Myanna Lahsen, Manjana Milkoreit, Berta Martin López, Kimberly A. Nicholas, Rika Preiser, Gaia Vince, Joost M. Vervoort, and Xu. Jianchu. 2016. Bright spots: Seeds of a good Anthropocene. *Frontiers in Ecology and the Environment* 14:441–448.

Boulton, Elizabeth. 2016. Climate change as a 'hyperobject': A critical review of Timothy Morton's reframing narrative. *Wiley Interdisciplinary Reviews. Climate Change* 7:772–785. https://doi.org/10.1002/wcc.410.

Brauch, Hans Günter. 2021. The Anthropocene concept in the natural and social sciences, the humanities and law – a bibliometric analysis and a qualitative interpretation (2000–2020). *In Paul J. Crutzen and the Anthropocene: A new epoch in Earth's history*, ed. Susanne Benner, Gregor Lax, Paul J. Crutzen, Ulrich Pöschl, Jos Lelieveld and Hans Günter Brauch, 289–438. Cham: Springer.

Brossard, Dominique. 2013. New media landscapes and the science information consumer. *Proceedings of the National Academy of Sciences of the United States of America* 110:14096–14101. https://doi.org/10.1073/pnas.1212744110.

Bucchi, Massimiano, and Brian Trench. 2014. *Routledge handbook of public communication of science and technology*. London: Routledge. https://doi.org/10.4324/9780203483794.

Carver, Rebecca Bruu. 2014. Public communication from research institutes: Is it science communication or public relations? *JCOM. Journal of Science Communication* 13:C01. 10.22323/2.13030301.

Castree, Noel. 2015. Changing the Anthropo(s)cene: Geographers, global environmental change and the politics of knowledge. *Dialogues in Human Geography* 5:301–316. https://doi.org/10.1177/2043820615613216.

Ceballos, Gerardo, Paul R. Ehrlich, and Rodolfo Dirzo. 2017. Biological annihilation via the ongoing sixth mass extinction signaled by vertebrate population losses and declines. *Proceedings of the National Academy of Sciences* 114 (30): E6089–E6096. https://doi.org/10.1073/pnas.1704949114

Chakrabarty, Dipesh. 2009. The climate of history: Four theses. *Critical Inquiry* 35:197–222. https://doi.org/10.1086/596640.

———. 2021. *The climate of history in a planetary age*. Chicago: University of Chicago Press. https://doi.org/10.7208/chicago/9780226733050.001.0001.

Chaplin, Joyce E. 2023. Historians of ideas rush in where Stratigraphers fear to tread. *Journal of the History of Ideas* 84:775–801. https://doi.org/10.1353/jhi.2023.a909538.

Collins, Kimberley, David S. Shiffman, and Jenny Rock. 2016. How are scientists using social Media in the Workplace? *PLoS One* 11:e0162680. https://doi.org/10.1371/journal.pone.0162680.

Damianos, Alexander. 2021. Ratifying the Anthropocene: A study of the Anthropocene working Group's ongoing effort to formalize the Anthropocene as a unit of the geologic time scale. PhD thesis. London School of Economics and Political Science.

Darling, Emily S., David S. Shiffman, Isabelle M. Côté, and Joshua Adam Drew. 2013. The role of Twitter in the life cycle of a scientific publication. *Ideas in Ecology and Evolution* 6:32–43. https://doi.org/10.4033/iee.2013.6.6.f.

Dickens, Peter. 2015. Leaving the world instead of saving it. *Capitalism Nature Socialism* 26 (4): 249–252. https://doi.org/10.1080/10455752.2015.1069480.

Dunwoody, Sharon. 1999. Scientists, journalists, and the meaning of uncertainty. In *Communicating uncertainty: Media coverage of new and controversial science*, ed. Sharon M. Friedman, Sharon Dunwoody, and Carol L. Rogers, 59–79. New York: Routledge. https://doi.org/10.4324/9781410601360-10.

Dürbeck, Gabriele, and Philip Hüpkes. 2020. *The Anthropocenic turn*. New York: Routledge. https://doi.org/10.4324/9781003037620.

Ellis, Erle C. 2018. *Anthropocene. A very short introduction*. Oxford: Oxford University Press.

Evans, Meredith, and Nadine Ryan. 2019. (De)compositions: A review of "Anthropocene". Visual and new media review. *Fieldsights*. https://culanth.org/fieldsights/decompositions-a-review-of-anthropocene. Accessed 29 Apr 2024.

Fay, Jennifer. 2018. *Inhospitable world: Cinema in the time of the Anthropocene*. New York: Oxford University Press. https://doi.org/10.1093/oso/9780190696771.001.0001.

Finney, Stanley C., and Lucy E. Edwards. 2016. The "Anthropocene" epoch: Scientific decision or political statement? *GSA Today* 26:4–10. https://doi.org/10.1130/gsatg270a.1.

Forman, Paul. 1971. Weimar culture, causality, and quantum theory, 1918–1927: Adaptation by German physicists and mathematicians to a hostile intellectual environment. *Historical Studies in the Physical Sciences* 3:1–115. https://doi.org/10.2307/27757315.

Friedman, Sharon M., Sharon Dunwoody, and Carol L. Rogers. 1999. *Communication uncertainty: Media coverage of new and controversial science*. Mahwah, NJ: Lawrence Erlbaum Associates.

Haff, Peter. 2024. Being human in the Anthropocene. https://blogs.nicholas.duke.edu/anthropocene/. Accessed 25 May 2024.

Hamilton, Clive. 2017. *Defiant earth: The fate of humans in the Anthropocene*. Cambridge: Polity.

Hara, Noriko, Jessica Abbazio, and Kathryn Perkins. 2019. An emerging form of public engagement with science: Ask me anything (AMA) sessions on Reddit r/science. *PLoS One* 14:e0216789. https://doi.org/10.1371/journal.pone.0216789.

Hart, P. Sol, and Anthony Leiserowitz. 2009. Finding the teachable moment: An analysis of information-seeking behavior on global warming related websites during the release of the day after tomorrow. *Environmental Communication* 3:355–366. https://doi.org/10.1080/17524030903265823.

Haus der Kulturen der Welt. 2022. The Anthropocene at HKW - From 2013–2022. https://archiv.hkw.de/en/programm/themen/das_anthropozaen_am_hkw/das_anthropozaen_am_hkw_start.php. Accessed 15 May 2024.

Heise, Ursula K., Jon Christensen, and Michelle Niemann, eds. 2017. *The Routledge companion to the environmental humanities*. London, New York: Routledge.

Holzer, Boris, and Leslie Sklair. 2021. Western Europe: Planetary eurocentrism. In *The Anthropocene in global media*. Neutralizing the risk, ed. Leslie Sklair, 159–186. Abingdon: Routledge.

Huang, Ying, and Weishan Miao. 2021. Re-domesticating social media when it becomes disruptive: Evidence from China's "super app" WeChat. *Mobile Media & Communication* 9 (2): 177–194. https://doi.org/10.1177/2050157920940765.

Jagodzinski. 2018. *Interrogating the Anthropocene: Ecology, aesthetics, pedagogy, and the future in question*. Cham: Springer. https://doi.org/10.1007/978-3-319-78747-3.

Jamieson, Kathleen Hall, Dan M. Kahan, and Dietram A. Scheufele. 2017. *The Oxford Handbook of the Science of Science Communication*. Oxford: Oxford University Press.

Johnson, Keith Leslie. 2014. Manga in the Anthropocene: Notes toward a cyberpunk ecology 2. *Southeast Review of Asian Studies* 36:111–122.

Johns-Putra, Adeline. 2010. Ecocriticism, genre, and climate change: Reading the utopian vision of Kim Stanley Robinson's science in the capital trilogy. *English Studies* 91 (7): 744–760. https://doi.org/10.1080/0013838X.2010.518043.

———. 2016. Climate change in literature and literary studies: From cli-fi, climate change theater and ecopoetry to ecocriticism and climate change criticism. *Wiley Interdisciplinary Reviews: Climate Change* 7 (2): 266–282. https://doi.org/10.1002/wcc.385.

Jørgensen, Finn Arne, and Dolores Marie Jørgensen. 2016. The Anthropocene as a history of technology. *Technology and Culture* 57:231–237. https://doi.org/10.1353/tech.2016.0026.

Kember, Sarah. 2017. After the Anthropocene: The photographic for earthly survival? *Digital Creativity* 28:348–353. https://doi.org/10.1080/14626268.2017.1380048.

Keogh, Luke, and Nina Möllers. 2014. Pushing boundaries: Curating the Anthropocene at the Deutsches museum, Munich. In *Climate change and museum futures*, ed. Fiona Cameron and Brett Neilson, 90–101. New York: Routledge. https://doi.org/10.4324/9780203752975-11.

Lewis, Simon L., and Mark A. Maslin. 2018. *The human planet. How we created the Anthropocene*. New Haven, CT: Yale University Press.

Lorimer, Jamie. 2017. The Anthropo-scene: A guide for the perplexed. *Social Studies of Science* 47:117–142.

Lyons, Siobhan. 2017. Extinction or survival: How storytellers explore the ethics of colonising other planets. *The Conversation*. https://theconversation.com/extinction-or-survival-how-storytellers-explore-the-ethics-of-colonising-other-planets-73595. Accessed 15 May 2024.

Malm, Andreas. 2016. *Fossil capital. The rise of steam power and the roots of global warming*. London: Verso.

Marcinkowski, Frank, Matthias Kohring, Silke Fürst, and Andres Friedrichsmeier. 2014. Organizational influence on scientists' efforts to go public. *Science Communication* 36 (1): 56–80. https://doi.org/10.1177/1075547013494022

Matthews, John A. 2019. Book review of Zalasiewicz, J., Waters, C.N, et al., eds., The Anthropocene as a geological time unit, CUP 2019. *Holocene* 29 (8): 1396–1397.

Milner, Andrew J., ed. 2018. *Again, dangerous visions. Essays in cultural materialism*. Leiden, Boston: Brill.

Möllers, Nina, Christian Schwägerl, and Helmut Trischler. 2015. *Welcome to the Anthropocene*. München: Deutsches Museum.

Moore, Jason W., ed. 2016. *Anthropocene or Capitalocene? Nature, history, and the crisis of capitalism*. Oakland, CA: PM Press.

Morton, Timothy. 2013. *Hyperobjects: Philosophy and ecology after the end of the world*. Minneapolis: University of Minnesota Press.

Nature - Editorial. 2011. The human epoch. *Nature* 473 (7347): 254. https://doi.org/10.1038/473254a.

Oldfield, Frank, Anthony D. Barnosky, John A. Dearing, Marina Fischer-Kowalski, John McNeill, Will Steffen, and Jan Zalasiewicz. 2013. The Anthropocene review: Its significance, implications and the rationale for a new transdisciplinary journal. *The Anthropocene Review* 1:3–7. https://doi.org/10.1177/2053019613500445.

Parikka, Jussi. 2015. The Anthropocene has been shaped by the media and our digital lives. *The Conversation*. 4 February 2015, https://theconversation.com/the-anthropocene-has-been-shaped-by-the-media-and-our-digital-lives-37124.

Peters, Hans Peter, and Sharon Dunwoody. 2016. Scientific uncertainty in media content: Introduction to this special issue. *Public Understanding of Science* 25 (8): 893–908. https://doi.org/10.1177/0963662516670765.

Pflugfelder, Ehren Helmut, and Alexander Mahmou-Werndli. 2021. Impacts of genre and access on science discussions: 'The New Reddit Journal of Science'. *Journal of Science Communication* 20 (05): A04. https://doi.org/10.22323/2.20050204.

Powell, Chloe Louise. 2017. Cli-Fi cinema: An epideictic rhetoric of blame. Master's thesis, University of Nevada, Las Vegas.

Rahmstorf, Stefan. 2004. The day after tomorrow—some comments on the movie. *Pik-Potsdam*. https://web.archive.org/web/20041011172259/http:/www.pik-potsdam.de/~stefan/tdat_review.html. Accessed 15 May 2024.

Raworth, Kate. 2014. Must the Anthropocene be a Manthropocene? *The Guardian*. https://www.theguardian.com/commentisfree/2014/oct/20/anthropocene-working-group-science-gender-bias. Accessed 15 May 2024.

Rea, Naomi. 2019. How teen climate activist Greta Thunberg inspired the art world to address the biggest problem facing humanity. *Artnet News*. https://news.artnet.com/art-world/greta-thunberg-climate-art-1645336. Accessed 15 May.

Reusswig, Fritz, and Anthony Leiserowitz. 2005. The international impact of the day after tomorrow. *Environment* 47:41–44. https://doi.org/10.1080/00139157.2005.10524447.

Ritchie, Stuart. 2020. *Science fictions: Exposing fraud, bias, negligence and hype in science*. London: Bodley Head.

Robin, Libby, and Cameron Muir. 2015. Slamming the anthropocene: Performing climate change in museums. *reCollections* 10: .

Robin, Libby, Dag Avango, Luke Keogh, Nina Möllers, Bernd Scherer, and Helmuth Trischler. 2014. Three galleries of the Anthropocene. *The Anthropocene Review* 1:207–224. https://doi.org/10.1177/2053019614550533.

Rosol C, Nelson S, Renn J. 2017. Introduction: In the machine room of the Anthropocene. *The Anthropocene Review* 4 (1):2–8. https://doi.org/10.1177/2053019617701165.

Salmose, Niklas. 2018. The apocalyptic sublime: Anthropocene representation and environmental agency in Hollywood action-adventure Cli-Fi films. *The Journal of Popular Culture* 51:1415–1433. https://doi.org/10.1111/jpcu.12742.

Sklair, Leslie. 2021. Media coverage of Anthropocene-related creative arts. In *The Anthropocene in global media. Neutralizing the risk*, ed. Leslie Sklair, 232–251. Abingdon: Routledge.

Sklair, Leslie, Chad Steacy, Jonathan DeVore, and Ron Wagler. 2021. The Anthropocene in North American media: The pursuit of the 'good' Anthropocene. In *The Anthropocene in global media. Neutralizing the risk*, ed. Leslie Sklair, 55–96. Abingdon: Routledge.

Steffen, Will, Jacques Grinevald, Paul Crutzen, and John McNeill. 2011. The Anthropocene: conceptual and historical perspectives. *Philosophical Transactions of the Royal Society A* 369 (1938): 842–867. https://doi.org/10.1098/rsta.2010.0327

Svoboda, Michael. 2014. A review of climate fiction (Cli-Fi) cinema … past and present. *Yale Climate Connections*.

Swanson, Heather Anne, Nils Bubandt, and Anna Lowenhaupt Tsing. 2015. Less than one but more than many: Anthropocene as science fiction and scholarship-in-the-making. *Environment and Society* 6. https://doi.org/10.3167/ares.2015.060109.

Trexler, Adam. 2015. *Anthropocene fictions: The novel in a time of climate change*. Charlottesville: University of Virginia Press. https://doi.org/10.5860/choice.193345.

Truscello, Michael. 2018. Catastrophism and its critics: On the new genre of environmentalist documentary film. In *Interrogating the Anthropocene*, ed. Jan Jagodzinski, 257–275. Cham: Palgrave Macmillan. https://doi.org/10.1007/978-3-319-78747-3_11.

Vidas, Davor, and Peter Johan Schei. 2013. *The World Ocean in globalisation: Climate change, sustainable fisheries, biodiversity, shipping, regional issues*. Leiden: Brill/Nijhoff. https://doi.org/10.1163/22116001-90000182.

Wang, Yiming, Qing Fang, and Wei Peng. 2018. China's recently launched English-language science and technology journals, 2012–16. *Journal of Scholarly Publishing* 50 (1): 37–47. https://doi.org/10.3138/jsp.50.1.07.

Wikipedia. 2024. Anthropocene Working Group. https://en.wikipedia.org/wiki/Anthropocene_Working_Group. Accessed 15 May 2024.

Weigelin-Schwiedrzik, Susanne. 2018. Doing things with numbers: Chinese approaches to the Anthropocene. *International Communication of Chinese Culture* 5 (1–2): 17–37. https://doi.org/10.1007/s40636-018-0115-8.

Wodak, Josh. 2017. Shifting baselines: Conveying climate change in popular music. *Environmental Communication* 12:58–70. https://doi.org/10.1080/17524032.2017.1371051.

Wride, Tim B. 2019. Earth works: Mapping the Anthropocene. *Norton Museum of Art*. https://www.justinbrice.com/earthworks-mapping-anthropocene. Accessed 15 May 2024.

Zalasiewicz, Jan, Colin N. Waters, Mark Williams, and Colin P. Summerhayes, eds. 2019. *The Anthropocene as a geological time unit. A guide to the scientific evidence and current debate*. Cambridge: Cambridge University Press.

Chapter 6
Conclusion: The Anthropocene Remade

The preceding chapters describe the Anthropocene Working Group's (AWG) effort to formalize the Anthropocene proposal from 2009 to 2023.

Beginning with a timeline of activities and events (Chap. 2), the subsequent Chap. 3 analyzes AWG members' scholarly activities and publications. While Chap. 4 focuses on the inter- and multidisciplinary ramifications of the Anthropocene concept, Chap. 5 provides an overview of the AWG's public outreach. Each of these analyses stands on its own regarding subject, methodology, and findings. This final chapter summarizes and synthesizes them to tease out the implications for the relationships between science and the public in the face of ecological and other problems that cannot be addressed without scientific expertise—but also cannot be solved by scientific knowledge alone.

Our analysis has portrayed the AWG as an example of contemporary science of geographically, topically, and culturally global scope. The intra-, inter-, transdisciplinary, and public debates around the AWG's topic, the geological Anthropocene, illustrate a multidimensional globalization of science. Addressing the Anthropocene entangles geological terms in a mesh of social and cultural notions, including how scholarly disciplines and media respond to Earth System Sciences (ESS) knowledge. Ancillary perspectives emerge in these contexts, such as the struggle of geoscience disciplines regarding their place in contemporary societies (Stewart et al. 2023).

Recognizing the AWG's significant role in advancing the Anthropocene concept, the analyses of this book focus on different aspects of the AWG's challenges and potential impact, ranging from intellectual and methodological aspects over navigating a bureaucratic science environment to mastering public outreach.

– *Analyses*: Chap. 2 centers on AWG's navigation within the geoscientific bureaucracy and potential changes to the Geological Time Scale (GTS). In contrast,

This chapter is authored by Martin Bohle, Boris Holzer, Leslie Sklair, Fabienne Will.

M. Bohle et al., *The Anthropocene Working Group and the Global Debate Around a New Geological Epoch*, Emerging Globalities and Civilizational Perspectives, https://doi.org/10.1007/978-3-031-85175-9_6

Chap. 3 describes the scholarly activities of the AWG members, which consisted of coauthored publications but also involved continuous individual contributions to the geosciences and other disciplines. Chapter 4 focuses on the Anthropocene's conceptual complexity and its implications for interdisciplinary research. Chapter 5 highlights the aspects of the AWG's work's outreach and societal engagement.

- *Challenges*: Chaps. 2 and 5 highlight challenges related to the AWG's position within the geoscientific community and the public. Chapters 3 and 4 emphasize intellectual and methodological challenges, particularly in addressing the Anthropocene as a scientific problem and integrating humanities and geosciences.
- *Impacts*: Chap. 2 addresses knowledge building within geosciences. While Chap. 3 traces the disciplinary impact of the AWG, Chap. 4 proposes the emergence of new interdisciplinary fields. Chapter 5 underscores the broad societal and cultural influence of the Anthropocene concept.

6.1 Mandate and Mission of the AWG

The AWG's mandate, given by the International Commission on Stratigraphy (ICS) and the International Union of Geological Sciences (IUGS), was to discuss a question of geological chronostratigraphy. The publicly perceived AWG mandate, however, was (and is) more about establishing a scientifically sound concept of the Anthropocene "across the board." This divergence of perceptions has occupied a substantial share of the Anthropocene literature, discussing, e.g., what the AWG is supposed to do and what the term Anthropocene implies (Zalasiewicz et al. 2021; Table 1, pp. 11–12). The AWG itself has maneuvered between these two poles, on the one hand reformulating the question of the Anthropocene to meet the established standards of the Geological Time Scale and, on the other hand, catering to the public interest and the mass media (Damianos 2024). A consequence of the diverse perceptions of the AWG's mandate was irritation, be it of academics (e.g., (Lundershausen 2018)) or officers of the ICS/IUGS, who asked: "Is the 'Anthropocene' a concept in search of a stratigraphic record" (Finney 2014; p. 23)? The irritation of ICS/IUGS officers and other parts of the geoscientific community contributed to the relatively straightforward rejection of the AWG proposal (Witze 2024a, b) by the Subcommission on Quaternary Stratigraphy (SQS) endorsed by the parent bodies ICS and IUGS (IUGS 2024).

How deep the irritation runs is illustrated by the reception of the proposal of an Anthropocene "episode" (Waters et al. 2022). This proposal looked like a possible intradisciplinary compromise. However, the publication has not triggered any noticeable reaction so far. For example, the detailed discussion in "The Anthropocene Event as a Cultural Zeitgeist in the Earth-Human Ecosystem" (Koster et al. 2024) does not discuss it. Many reports and publications between 2009 and 2024 illustrate divergent views regarding the AWG's mandate, which were the main driver of interdisciplinary debates about the ramifications and implications of (a geological notion of) an Anthropocene.

The AWG's scholarly composition and the analysis of its members' contributions illustrate a selection of expertise from a relatively diverse set of academic disciplines, although weighted toward geosciences. This diversity within the AWG has ascertained that different views on the notion of an Anthropocene could be debated. After the AWG voted in favor of an Anthropocene epoch in 2016 and 2019, the ICS/IUGS management insisted on distinguishing between voting and advisory members. This insistence indicated that the AWG membership seemed too diverse to inform a decision about the Geological Time Scale within the IUGS.

Although the AWG's composition reflected an increasingly globalized science, its membership was skewed toward Western (anglophone) societies. Considering the likely geographical spread of impacts of planetary-scale anthropogenic change (Kotz et al. 2024), the AWG composition seemed insufficient (Yusoff 2018). Despite its multidisciplinary and global outlook, the AWG's inclusiveness and diversity were limited, although they might have matched the degree of social and cultural inclusiveness of contemporary geosciences. Still, AWG members' scope of expertise, their engagement with the public, and their ability to liaise with cultural institutions (Rosol et al. 2023) to some degree compensated for its limited inclusiveness and diversity.

In retrospect, the AWG's name (The Working Group on the 'Anthropocene') and the compact version (The Anthropocene Working Group) facilitated an over-interpretation of the AWG's mandate, at least for non-geologists. Many observers also failed to grasp the dedicated meaning of the suffix '-cene in geology, i.e., which points to a particular feature of the GTS and a related methodological corpus, and instead, consider the term Anthropocene as an indicator of cultural change (Dalby 2015). Considering the immanent conflicts between a strictly geological concept of the Anthropocene and its much broader public and normative interpretations, the International Science Council rather than the IUGS could have been a more appropriate institutional and epistemic environment for an 'Anthropocene Working Group' aiming at a well-balanced *zeitdiagnostic* analysis, i.e., an analysis of THE' Anthropocene question. The AWG of the ICS/IUGS faced this expectation and, to some extent, accepted it as a vital part of its mission, but it was ultimately bound to the geological subquestion of a chronostratigraphic boundary. The AWG delivered according to its mandate, but its master did not receive the message well.

6.2 Negotiating Boundaries

On the one hand, the work of the AWG has caused methodological debates within geoscience communities, particularly concerning the clarity of geological terms that had evolved over extended periods. While terms such as epoch, event, or episode,[1] which also have an ordinary meaning, seem specific to geoscience, a closer look reveals areas of interpretation and choice (Waters et al. 2022; Merritts et al. 2023), requiring effort and time to consolidate a shared understanding. On the other hand, geological research requires much attention to detail, imagination, and diverse knowledge (Zalasiewicz 2010). These features of geology highlight a discipline-specific process of knowledge-building involving narratives and peer consensus (Phillips 2012). Hence, refining terminology through consensus is a tested geological practice, including a reluctance to alter the GTS because of the substantial sunken effort (Gradstein and Ogg 2020). Most intra-geology debates can be framed this way. Observers from other disciplines (and the public) seem less aware of knowledge-building processes in geology. Given the GTS's epistemic status and geologists' professional practices, Paul Crutzen's initial Anthropocene Proposal was very unconventional, pioneering the "essentially political valence of the term" (Damianos 2024) [p. 2].

Although debates of AWG members with peers (Zalasiewicz et al. 2019) have elaborated an appropriate geological terminology (e.g., epoch, event, episode) and corresponding evidence practices (e.g., hard/soft strata, kind of marker), they have also included cultural perspectives, such as a stasis of geological processes, distinct segregation of human and natural worlds, and philosophical concerns of preserving hegemonic guidance of geology's insights into "deep time" (Bjornerud 2018). Against the backdrop, the AWG's proposal contrasted with the disciplinary-cultural backgrounds of many geologists. Hence, the AWG proposal might have been too swift and brutal, shaking established worldviews concerning nature, earth, and the world, including the habitual segregation of historical and geological time scales. Many geoscientists also objected to the AWG's blurring of the boundary between scientific fact-finding and public outreach, regarding it as an illegitimate effort to 'politicize' science. Given that the AWG proposal demarcates a historical cusp,[2] many geologists at least might have welcomed a less prominent and clearly demarcated, diachronically more flexible geological concept like "event." However, taking this route carries a scent of denial of the principal nature of how anthropogenic impact may alter Earth System dynamics qualitatively.

---

[1] Interestingly, the term "Renaissance" (a multifaceted period of change in European post-medieval times) was invoked (Koster et al. 2023, 2024) arguing that the Anthropocene is an event and not an epoch. However, this argument should consider that European Renaissance states embarked on the global projection of their power (Mokyr 2017; Reinhard 2016); therefore, associating the question of "who powered the Anthropocene" might be permitted when considering it Renaissance-like.

[2] Given the Great Acceleration, the Earth System's dynamics diverge from those characterizing the Holocene.

6.3 Within and Beyond the Geo- and Earth Sciences

The interplay between the AWG and other bodies of the international geological community has been complicated. Officially established in 2009, the AWG presented its findings in 2016 at the 35th International Geological Congress (IGC). The management of the ICS and the IUGS halted this attempt, considering it premature and, thus, disappointing substantial public expectations by overseeing the matter behind closed doors. Subsequently, the ICS and the IUGS insisted on the 'chronostratigraphic mandate,' reconducted the AWG, and re-composed its membership after the AWG had confirmed its initial 2016 findings in 2019. In 2024, the Anthropocene concepts were officially debated at the 37th IGC. A few months before, on October 31, 2023, the AWG had submitted a formal proposal to the SQS, the next highest level in the ICS/IUGS system. The AWG proposed to amend the GTS by adding a novel epoch, the Anthropocene (Waters et al. 2024a, b). Following its 'chronostratigraphic mandate' and relegating *zeitdiagnostic* elements of analysis to the margin, the AWG presented a Global Boundary Stratotype Section and Point (GSSP) using modern chronostratigraphic methods. The GSSP demarcates the lower boundary of the Anthropocene in the GTS in the mid-twentieth century. The required isochronous signal is situated at the historical time of the 'Great Acceleration' when increasing resource consumption of affluent Western societies began to shift the physical and biological dynamics of the Earth System. In line with its mandate, the AWG employed an interdisciplinary methodology that amalgamated Earth System Science (ESS), geological stratigraphy, and contemporary observational techniques.[3] The AWG argued that the Anthropocene represents a legitimate choice of a novel geological epoch suitable for inclusion in the GTS, terminating the current geological epoch, the Holocene. The proposed 'Golden Spike' for the Anthropocene, or GSSP, would be the first GTS feature attributable to (collective) human action. The following less recent GSSP, the boundary of the Meghalayan stage of the Holocene (4.200 years BP), is attributed to natural climatic change (drought) impacting early human civilizations in Eastern Mediterranean, Mesopotamia, Indus Valley, and Yangtze River Valley.

The SQS rejected the AWG proposal at the beginning of March 2024. The ICS/IUGS management initiated the decision process with the consent of one (of three) SQS chairpersons (Witze 2024a). The decision was leaked (Ansede 2024) and confirmed by the IUGS (Witze 2024b). About 3 weeks later, the ICS/IUGS management published, on March 20, 2024, an account of the decision (IUGS 2024). Going public with a chronostratigraphic decision represented a paramount shift in IUGS's communication practices.

The IUGS statement describes how the Anthropocene notion emerged, how the AWG handled it, that the AWG's proposal was rejected, why the IUGS prefers the

---

[3] First, the AWG used ESS findings, namely that human activities associated with the Great Acceleration, drive a state shift in Earth System dynamics. Second, the AWG considered that radioactive fallout could mark this shift and serve as GSSP to label the onset of the Anthropocene.

concept of an Anthropocene event, and future uses of the Anthropocene notion. It emphasizes the origin and development of the Anthropocene notion partly outside geology, compliance with procedures of the ICS/IUGS system, time-transgressive and spatially and temporally variable human effects on global systems, and the usefulness of the term Anthropocene as an invaluable descriptor in human-environment interactions. The AWG members' immediate reaction (Turner et al. 2024; Zalasiewicz 2024; Barnosky and Hannibal 2024; Horn and Wagreich 2024) reiterates what their opponents in ICS/IUGS decline, insisting on the shift of the dynamics of the Earth System (Steffen et al. 2018; Summerhayes et al. 2024). Regarding a shift of the Earth System because (IUGS 2024, pp. 1–2) "human activity was impacting natural environmental conditions to the extent that we had effectively left the natural stable conditions of the Holocene," the IUGS argues in analogy with "the great transformative events in Earth history such as the Great Oxygenation (2.4-2.1 Ga), the Cambrian Explosion, or the Great Ordovician Biodiversification events" that "[n]one of these major transformative events in Earth history are represented as chronostratigraphic units, and hence there has been no requirement for formal ratification." The IUGS prefers the notion of a geological event instead of a geological epoch because it captures human activity's diverse, regional, and diachronous impacts on natural environments. It states that professional practices (of the GTS) describing system shifts in deep time should also apply to ongoing planetary-scale anthropogenic change. However, the disapproval of *zeitdiagnostic* elements is not shared by the entire geoscientific community. For instance, it contrasts with the views of the IUGS president (2020–2024), John Ludden, who wrote in his message[4] to the 37th IGC, "As geologists, we have the Anthropocene – I believe that apart from the word Jurassic (as in Jurassic Park), this is one of the few geological terms that the public understands – they acknowledge that as humankind our effects on the planet are measurable and long-lasting. We as geoscientists are somewhat backward-looking in that we obsess over when the Anthropocene started – we need to embrace the concept as it is beyond the realms of traditional geology and stratigraphy – it is the geology of humanity."

In hindsight, ESS has been instrumental in shaping the initial Anthropocene concept and its formalization in geological terminology. Originating within the ESS community, the term Anthropocene resonates with various scholarly disciplines, including geology. Therefore, it raises questions about scientific authority and epistemological leadership. Likely, it was of little help that the interacting geoscience disciplines belong to different ladders of international scientific organizations. Furthermore, these disciplines differ substantially in core methodologies, for example, in mathematization, experimental control, reproducibility, and application of simulation tools. Hence, communication between the geoscience disciplines ESS and geology is demanding. In particular, qualifying a change in the Earth System dynamics as a 'state shift,' 'phase shift,' or 'tipping point' exercised by human activity (as done in ESS (Rockström et al. 2024)) implies *zeitdiagnostic*

[4] https://www.igc2024korea.org/content/14280.

connotations, which many geologists might not be willing to follow (yet) because of their philosophical positions (Hamilton 2016; Bohle 2023; Koster et al. 2024; Barnosky and Hannibal 2024). Agreeing on the Anthropocene epoch would manifest human prowess as a novel transformative force along with the geological processes of the Earth system, which would imply transcending geology's traditional boundaries of academic insights and disciplinary worldviews. Early in the Anthropocene debates, a geologist's irritation was phrased, "Is the Anthropocene an issue of stratigraphy or pop culture?" (Autin and Holbrook 2012, p. 60). To which the AWG responded:

> "The concept has certainly gone beyond the confines of stratigraphic research. But much of the interest has been among the wider scientific community, because Anthropocene explicitly compares human perturbation of the Earth system (as observed) with ancient natural perturbations (as preserved via proxy evidence in rock strata), and it considers together and integrates diverse forms of environmental change. This has a novel (in stratigraphy) corollary that the wider scientific community should be considered….
>
> We regard broader popular interest as positive. The Anthropocene has provided a longer-term perspective of humanity's activities and brings stratigraphic principles and practice to a broader audience. Also, the phenomenon of contemporary global change—perhaps unlike the formal determination of past geological time units—potentially concerns everyone". (Zalasiewicz et al. 2012, p. e21)

While the formal endorsement of the Anthropocene as a geological epoch was aborted in 2024, the AWG's deliberations still invite reflection across geological, philosophical, and political fields. The history of the AWG's dealings in cooperation with cultural institutions witnesses these reflections. With the Anthropocene concept, different time scales, which have been perceived as distinct for the last 250 years, converge in novel ways. The geo-cultural debate on the Anthropocene thus has stimulated a shift in how different disciplines conceptualize time, combining deep with historical time and the local with the planetary.

6.4 Disciplinary Autonomy and Multidisciplinary Debates

Challenging disciplinary boundaries and classical narratives and stimulating philosophical and cultural reconsiderations at different scales from the local to the planetary, the Anthropocene concept has attracted scholarly attention and stirred controversy within Earth sciences and the social sciences, environmental humanities, and the creative arts. The decision to regard the Anthropocene question as a chronostratigraphic one falling within the disciplinary autonomy of geoscience constituted but also limited the epistemic authority of the AWG to actually 'decide' the Anthropocene question. As the discussion at the 35th IGC determined, the AWG had no choice but to align with the disciplinary chronostratigraphic views of the IUGS/ICS. During the subsequent years, the AWG delivered according to its mandate, seizing the opportunity offered by the cooperation with the *Haus der Kulturen der Welt* (Berlin). The rejection by the ICS/IUGS is not an end to discussing planetary-scale anthropogenic change among geologists and their peers. A firmly established description of the geo-cultural 'present,' the Anthropocene, is available, including its geological attributes.

Public discourse has been based on a broad interpretation of the Anthropocene and its consequences for a long time, making any significant impact of the rejection unlikely. In contradiction to the AWG's mandate to establish a chronostratigraphic indicator (GSSP) and, thus, to locate the beginning of the Anthropocene in the past, public and interdisciplinary debates, as well as cultural imaginaries, invoke the Anthropocene to problematize the future.

Initially driven by some earth scientists and geologists, today's Anthropocene discourse is neither dependent on nor determined by geosciences anymore. Social sciences and environmental humanities scholars have adopted and popularized the concept of an 'Anthropo-scene.' Despite interdisciplinary dialogs, that discourse has not produced consensus but instead highlighted the multiple perspectives on the term (Thomas et al. 2020).

While the specific geological concept elaborated by the AWG may be accessible to empirical investigation, it appeared to be too broad (and probably too political) for the IUGS and too narrow for many observers outside the geoscientific community. Given that the ICS/IUGS is a modest public player different from, for example, the WMO (World Meteorological Organization), that organization and the ESS community are more plausible candidates to drive the future discussion about the stage shift of the Earth System dynamics centered on climate change and biodiversity loss (Rockström et al. 2024; Kotz et al. 2024). The IUGS decision will likely not significantly alter the future Anthropocene discourse. The reaction triggered by the rejection decision indicates that it fueled rather than tamed the debate. Yet, the rejection by the ICS/IUGS lends itself to an interpretation 'of denial' of anthropogenic global change that will find its audience. In any case, the accelerated opening of geoscience disciplines for interaction with other disciplines, as the advocates of the event concept advocate (Koster et al. 2024), would be a favorable intermediate development.

The AWG proposal implies a distinct *zeitdiagnostic* meaning. The onset of the Anthropocene is equivalent to the end of the Holocene, i.e., the present interglacial period characterized by a given variability of characteristics of the Earth System. When displaying natural dynamics without human impact, the current interglacial (the Holocene) should give way to the next glacial period. The opposite, namely, ending the current interglacial by substantially warming the planet because of human activities, is a robust diagnostic of contemporary times (Steffen et al. 2018; Summerhayes et al. 2024) and the impact of human institutions and politics of the Holocene (Biermann et al. 2012; Dryzek and Pickering 2018; Wallenhorst 2023). Hence, the *zeitdiagnostic* Anthropocene question is addressed by what the former (and dissenting) member of the AWG, Erle Ellis, qualifies as a "positive feedback loop of 'runaway sociocultural niche construction'" (Ellis 2024, p. 1). The AWG considers the Great Acceleration (of Western-society-like affluence) the central feature of this runaway process leading to a stage shift of Earth System dynamics. Described in these terms, it seems appropriate to use nuclear fallout as the

Anthropocene's GSSP feature, given that a 'runaway nuclear reaction'" leads to the detonation of a nuclear weapon.

While the status of the Anthropocene within the geological terminology remains ambiguous, AWG's proposition has already precipitated a paradigmatic shift across geological, philosophical, and political domains. The conceptual framework of an Anthropocene onsetting with the Great Acceleration aids in conceptualizing the seismic shifts associated with Earth's and global society's transition beyond the epoch that (currently) geology calls the Holocene, including the need to formulate innovative policy strategies.

6.5 The Anthropocene Reconsidered

The AWG was a peculiar case of organizing science, gaining visibility through the promoter Paul Crutzen (Atmospheric Chemist and Nobel Laureate), who suggested amending the Geological Time Scale by ending the Holocene (Crutzen 2002). Studying Crutzen's Anthropocene Proposal within the limited professional scope of the ICS and IUGS accentuated the intellectual and organizational challenges for the AWG, including how to manage, in these circumstances, the interconnectedness of scientific and "other" fields of knowledge. To the IUGS's surprise, the AWG's studies gave rise to a multifaceted discourse on specific terminologies and disciplinary orientations, the Anthropocene concept's societal relevance, and its wider cultural ramifications. Finally, the AWG became the focal point of an international multidisciplinary debate straddling the boundaries between academic and public perceptions of science, including profound interaction with protagonists of culture and arts (Rosol et al. 2023).

In executing a scientific mandate of limited scope, the AWG amalgamated findings of ESS, geological stratigraphy practices, and contemporary observational techniques. The latter two choices were conservative within the corpus of established geological methodologies; the former was innovative. The AWG's main finding (Waters et al. 2016) was that the dynamics of the Earth System had shifted at a moment in historical time, confirming the end of the Holocene as initially proposed by Paul Crutzen (Crutzen and Stoermer 2000). The second vital finding was the insight that the shift of the Earth System happened in response to the expanding affluence of human consumption (McNeill and Engelke 2014). The third and crucial step was to seek (as requested by geological methodologies) isochronous planetary (chemical) signals (e.g., the fallout of Plutonium isotope from testing of nuclear weapons in the atmosphere) to mark potential chronostratigraphic boundaries. The fourth and innovative thought was to notice the occurrence of the 'shift of Earth System dynamics,' the 'isochronous planetary chemical signals,' and the 'expanding affluence,' all happened in the same period, the mid-twentieth century. Subsequently, it was a (good geological) practice to propose the related GSSP, or 'Golden Spike' in geology's metaphorical language (Waters et al. 2023).

Bound by its specific mandate, supervised by the geological establishment represented by the ICS and IUGS, the AWG did not explicitly consider broader cultural and societal questions of Crutzen's Anthropocene Proposal. However, several members addressed these issues, e.g., see in (Thomas 2022). Therefore, it can be assumed that the AWG members were well aware of the implications of their findings, namely that the Great Acceleration (mainly in Western societies) of expanding human affluence triggered the transition of the Earth System from the Holocene to the Anthropocene (Zalasiewicz et al. 2024; Turner et al. 2024). Likewise, they had to realize the associated political and cultural implications, including the likely reaction of the geological establishment represented by officers of the ICS and IUGS, who had been critical of the idea of an Anthropocene (Finney 2014; Walker et al. 2024).

Despite rejecting the AWG proposal, the ICS/IUGS management did not question the scientific core of the AWG's proposal (Turner et al. 2024; Zalasiewicz 2024), the qualitative shift of the Earth System dynamics. This specific attribute indicates issues of political and cultural choice (Walker et al. 2024). The ICS/IUGS position combines an epistemically sound (although conservative) stance with several 'shades of denial,' from denying that non-consolidated lake sediments (instead of rock and stone) qualify as a stratigraphic reference site to denying the end of typical dynamics of the Holocene. Another criticism (Koster et al. 2023; 2024) concerns the use of radioactive fallout from testing nuclear weapons as a reference point. Instead, evolving socio-ecological transformations are considered to shape a geological event of a human niche building (Ellis 2024). The latter view easily accommodates philosophies like ecomodernism (Ellis 2011). Bundling these views with the power of senior officers of the ICS (Ph. Gibbard) and IUGS (S. Finney) consolidated a hegemonic position unfavorable for amending the GTS.

In conclusion, the AWG served as a prism of the contested boundaries between science and society in the face of global existential problems, highlighting the interplay among diverse academic disciplines and the transformative impact of the Anthropocene concept.

The AWG has provided a laboratory for turning the important but rather general idea of 'planetary-scale anthropogenic change' into a precise geological concept. Based on years of intense and interdisciplinary work that had both popularized the term and established it as a research theme, it presented a proposal respecting disciplinary obligations for amending the GTS. For some AWG members, the formal rejection of the proposal through the IUGS might be an annoying setback, which highlights to a lesser degree a precarious epistemic status of the Anthropocene concept but more the limited institutional influence of its protagonists. Despite the failed attempt to formally codify the onset of the Anthropocene as an epoch in the GTS, the scientific and public debate about the Anthropocene continues to unfold (Summerhayes et al. 2024), given its enormous consequences for nature and society.

References

Ansede, Manuel. 2024. *The Anthropocene war: The controversial vote that decided not to change the planet's geological epoch*. El Pais. https://english.elpais.com/science-tech/2024-03-08/the-anthropocene-war-the-controversial-vote-that-decided-not-to-change-the-planets-geological-epoch.html# (accessed 11th March 2024)

Autin, Whitney J., and John M. Holbrook. 2012. Is the Anthropocene an issue of stratigraphy or pop culture? *GSA Today* 22:60–61. https://doi.org/10.1130/G153GW.1.

Barnosky, Anthony D., and Ellen Hannibal. 2024. Despite official vote, the evidence of the Anthropocene is clear. *Yale Environment*360. https://english.elpais.com/science-tech/2024-03-08/the-anthropocene-war-the-controversial-vote-that-decided-not-to-change--the-planets-geological-epoch.html# (accessed 5th Dec. 2024)

Biermann, Frank, K. Abbott, S. Andresen, K. Backstrand, S. Bernstein, M. M. Betsill, H. Bulkeley, et al. 2012. Navigating the Anthropocene: Improving earth system governance. *Science* 335:1306–1307. https://doi.org/10.1126/science.1217255.

Bjornerud, Marcia. 2018. *Timefulness—how thinking like a geologist can help to save the world*. Princeton: Princeton University Press.

Bohle, Martin. 2023. Crafting attributes of a geological now. *Earth Science, Systems and Society* 3:1–13. https://doi.org/10.3389/esss.2023.10089.

Crutzen, Paul J., and Eugen F. Stoermer. 2000. The "Anthropocene". *The Global Change Newsletter* 41:17–18. http://www.igbp.net/download/18.316f18321323470177580001401/1376383088452/NL41.pdf

Crutzen, Paul J. 2002. Geology of mankind. *Nature* 415: 23. https://doi.org/10.1038/415023a

Dalby, Simon. 2015. Framing the Anthropocene: The good, the bad and the ugly. *The Anthropocene Review* 3:1–19. https://doi.org/10.1177/2053019615618681.

Damianos, Alexander. 2024. Anthropocene angst: Authentic geology and stratigraphic sincerity. *Social Studies of Science*. https://doi.org/10.1177/03063127241282309

Dryzek, John S., and Jonathan Pickering. 2018. *The politics of the Anthropocene*. Oxford: Oxford University Press. https://doi.org/10.1093/oso/9780198809616.001.0001.

Ellis, Erle C. 2011. The planet of no return human resilience on an artificial earth. *The Breakthrough Institute* 2:11–16.

———. 2024. The Anthropocene condition: Evolving through social–ecological transformations. *Philosophical Transactions of the Royal Society B: Biological Sciences*379. https://doi.org/10.1098/rstb.2022.0255.

Finney, Stanley C. 2014. The 'Anthropocene' as a ratified unit in the ICS international Chronostratigraphic chart: Fundamental issues that must be addressed by the task group. *Geological Society, London, Special Publications* 395:23–28. https://doi.org/10.1144/SP395.9.

Gradstein, F. M., and J. G. Ogg. 2020. The chronostratigraphic scale. In *Geologic time scale 2020*, 21–32. Elsevier. https://doi.org/10.1016/B978-0-12-824360-2.00002-4.

Hamilton, Clive. 2016. The Anthropocene as rupture. *The Anthropocene Review* 3:93–106. https://doi.org/10.1177/2053019616634741.

Horn, Eva, and Michael Wagreich. 2024. *VAN statement on the IUGS Anthropocene decision*. Vienna Anthropocene Network (VAN). https://anthropocene.univie.ac.at/news-events/detail/news/van-statement-on-the-iugs-anthropocene-decision/ (accessed 14th May 2024)

IUGS. 2024. The Anthropocene. https://www.iugs.org/_files/ugd/f1fc07_40d1a7ed58de458c9f8f24de5e739663.pdf (accessed 22nd March 2024)

Koster, Emlyn, Philip L. Gibbard, and Mark Maslin. 2023. Optimising the Anthropocene definition: An epistemological view with briefings on four 2022-23 conferences. *Episodes*1–12. https://doi.org/10.18814/epiiugs/2023/023005.

———. 2024. The Anthropocene event as a cultural zeitgeist in the earth-human ecosystem. *Journal of Geoethics and Social Geosciences* 1:1–41. https://doi.org/10.13127/jgsg-43.

Kotz, Maximilian, Anders Levermann, and Leonie Wenz. 2024. The economic commitment of climate change. *Nature* 628. Springer US:551–557. https://doi.org/10.1038/s41586-024-07219-0.

Lundershausen, Johannes. 2018. Marking the boundaries of stratigraphy: Is stratigraphy able and willing to define, describe and explain the Anthropocene? *Geo: Geography and Environment* 5:e00055. https://doi.org/10.1002/geo2.55.

McNeill, J. R., and Peter Engelke. 2014. *The great acceleration. An environmental history of the Anthropocene since 1945*. Cambridge, MA: Harvard University Press. https://doi.org/10.4159/9780674970731.

Merritts, Dorothy, Lucy E. Edwards, Erle Ellis, Michael Walker, Stanley Finney, Philip L. Gibbard, Jacquelyn L. Gill, et al. 2023. Response to Waters et al. (2022) The Anthropocene is complex. Defining it is not. Earth-Science Reviews 238. Elsevier B.V: 104340. https://doi.org/10.1016/j.earscirev.2023.104340.

Mokyr, Joel. 2017. *A culture of growth. The origins of the modern economy*. Princeton, NJ: Princeton University Press.

Phillips, Jonathan. 2012. Storytelling in earth sciences: The eight basic plots. *Earth-Science Reviews* 115. Elsevier B.V:153–162. https://doi.org/10.1016/j.earscirev.2012.09.005.

Reinhard, Wolfgang. 2016. *Die Unterwerfung der Welt. Globalgeschichte der europäischen Expansion 1415-2015*. München: C.H. Beck.

Rockström, Johan, Louis Kotzé, Svetlana Milutinović, Frank Biermann, Victor Brovkin, Jonathan Donges, Jonas Ebbesson, et al. 2024. The planetary commons: A new paradigm for safeguarding earth-regulating systems in the Anthropocene. *Proceedings of the National Academy of Sciences* 121:1–10. https://doi.org/10.1073/pnas.2301531121.

Rosol, Christoph, Georg N. Schäfer, Simon D. Turner, Colin N. Waters, Martin J. Head, Jan Zalasiewicz, Carlina Rossée, Jürgen Renn, Katrin Klingan, and Bernd M. Scherer. 2023. Evidence and experiment: Curating contexts of Anthropocene geology. *The Anthropocene Review* 10:330–339. https://doi.org/10.1177/20530196231165621.

Steffen, Will, Johan Rockström, Katherine Richardson, Timothy M. Lenton, Carl Folke, Diana Liverman, Colin P. Summerhayes, et al. 2018. Trajectories of the earth system in the Anthropocene. *Proceedings of the National Academy of Sciences of the United States of America* 115:8252–8259. https://doi.org/10.1073/pnas.1810141115.

Stewart, Iain, Maria Angela Capello, Hassina Mouri, Kombada Mhopjeni, and Munira Raji. 2023. Three horizons for future geoscience. *Earth Science, Systems and Society* 3:1–6. https://doi.org/10.3389/esss.2023.10079.

Summerhayes, C. P., J. Zalasiewicz, M. J. Head, J. Syvitski, A. D. Barnosky, A. Cearreta, B. Fiałkiewicz-Kozieł, et al. 2024. The future extent of the Anthropocene epoch: A synthesis. *Global and Planetary Change* 242:104568. https://doi.org/10.1016/j.gloplacha.2024.104568.

Thomas, Julia Adeney. 2022. In *Altered earth*, ed. Julia Adeney Thomas. Cambridge, UK: Cambridge University Press. https://doi.org/10.1017/9781009042369.

Thomas, Julia Adeney, Mark Williams, and Jan Zalasiewicz. 2020. *The Anthropocene a multidisciplinary approach*. Cambridge, UK: Polity Press.

Turner, Simon D., Collins Waters, Jan Zalasiewicz, and Martin J. Head. 2024. What the Anthropocene's critics overlook- and why it really should be a new geological epoch. *The Conversation*. https://theconversation.com/what-the-anthropocenes-critics-overlook-and-why-it-really-should-be-a-new-geological-epoch-225493 (accessed 14th March 2024)

Walker, Michael J. C., Andrew M. Bauer, Matthew Edgeworth, Erle C. Ellis, Stanley C. Finney, Philip L. Gibbard, and Mark Maslin. 2024. The Anthropocene is best understood as an ongoing, intensifying, diachronous event. *Boreas* 53:1–3. https://doi.org/10.1111/bor.12636.

Wallenhorst, Nathanaël. 2023. *A critical theory for the Anthropocene*, Anthropocene—humanities and social sciences. Cham: Springer. https://doi.org/10.1007/978-3-031-37738-9.

Waters, Colin N., Simon D. Turner, Jan Zalasiewicz, and Martin J. Head. 2023. Candidate sites and other reference sections for the global boundary Stratotype section and point of the Anthropocene series. *The Anthropocene Review*205301962211364. https://doi.org/10.1177/20530196221136422.

Waters, Colin N., Simon Turner, An Zhisheng, Anthony Barnosky, Alejandro Cearreta, Andrew Cundy, Ian Fairchild, et al. 2024a. The Anthropocene Epoch and Crawfordian Age: Proposals by the Anthropocene Working Group Part 1: Anthropocene Series/Epoch: Stratigraphic context and justification of rank. https://eartharxiv.org/repository/view/6954/ (accessed 17th April 2024)

———. 2024b. The Anthropocene Epoch and Crawfordian Age: Proposals by the Anthropocene Working Group Part 2: Descriptions of the proposed Crawford Lake GSSP and supporting SABSs. https://eartharxiv.org/repository/view/6963/ (accessed17th March 2024)

Waters, Colin N., Jan Zalasiewicz, Colin Summerhayes, et al. 2016. The Anthropocene is functionally and stratigraphically distinct from the Holocene. *Science* 351 (6269): aad2622. https://doi.org/10.1126/science.aad2622

Waters, Colin N., Mark Williams, Jan Zalasiewicz, Simon D. Turner, Anthony D. Barnosky, Martin J. Head, Scott L. Wing, et al. 2022. Epochs, events and episodes: Marking the geological impact of humans. *Earth-Science Reviews* 234:104171. https://doi.org/10.1016/j.earscirev.2022.104171.

Witze, Alexandra. 2024a. Geologists reject the Anthropocene as Earth's new epoch—after 15 years of debate. *Nature*. https://doi.org/10.1038/d41586-024-00675-8.

———. 2024b. It's final: The Anthropocene is not an epoch, despite protest over vote. *Nature*. https://doi.org/10.1038/d41586-024-00868-1.

Yusoff, Kathryn. 2018. *A billion black Anthropocenes or none*, Forerunner. The University of Minnesota Press.

Zalasiewicz, Jan 2010. *The planet in pebble: A journey into Earth's deep history*. Oxford, UK: Oxford University Press.

Zalasiewicz, Jan 2024. Why criticisms of the proposed Anthropocene epoch miss the point. *New Scientist*.

Zalasiewicz, Jan, Alejandrol Cearreta, Paul Crutzen, Erls Ellis, Michael Ellis, Jacques Grinevald, Clément Poirier, et al. 2012. Is the Anthropocene an issue of stratigraphy or pop culture?: COMMENT. *GSA Today* 21:e21–e22. https://doi.org/10.1130/GSATG162C.1.

Zalasiewicz, Jan, Martin J. Head, Colin N. Waters, Simon D. Turner, Peter K. Haff, Colin Summerhayes, Mark Williams, et al. 2024. The Anthropocene within the geological time scale: A response to fundamental questions. *Episodes* 47:65–83. https://doi.org/10.18814/epiiugs/2023/023025.

Zalasiewicz, Jan, Colin N. Waters, Erle C. Ellis, Martin J. Head, Davor Vidas, Will Steffen, Julia Adeney Thomas, et al. 2021. The Anthropocene: Comparing its meaning in geology (Chronostratigraphy) with conceptual approaches arising in other disciplines. *Earth's Future* 9. https://doi.org/10.1029/2020EF001896.

Zalasiewicz, Jan, Colin N. Waters, Mark Williams, and Colin Summerhayes. 2019. In *The Anthropocene as a geological time unit*, ed. Jan Zalasiewicz, Colin N. Waters, Mark Williams, and Colin P. Summerhayes. Cambridge University Press. https://doi.org/10.1017/9781108621359.

Appendix

Annexe listing:

- The hyperlinks to documents (newsletters, annual/regular reports and files) that have been consulted (and downloaded) for the analysis presented in Chapters #2 #5; hyperlinks were consulted again on 31 Jan 2023 and 21 Jun 2023
- The membership of the Anthropocene Working Group (Table A1, A2, and A3).

M. Bohle et al., *The Anthropocene Working Group and the Global Debate Around a New Geological Epoch*, Emerging Globalities and Civilizational Perspectives, https://doi.org/10.1007/978-3-031-85175-9

Table A1 lists the newsletters of the Anthropocene Working Group (AWG). They appeared annually with exceptions. The original specific and master locations are given. The latter was updated (19th February 2025) during proofreading: http://quaternary.stratigraphy.org/working-groups/anthropocene, given that the site had been reorganised

Organisation	Document Type	Reference Period	Original Specific Location (hyperlink)	Master Location (hyperlink)
AWG	Newsletter	2009	http://quaternary.stratigraphy.org/wp-content/uploads/2018/08/Anthropocene-Working-Group-Newsletter-No1-2009.pdf	http://quaternary.stratigraphy.org/workinggroups/anthropocene
AWG	Newsletter	2010	http://quaternary.stratigraphy.org/wp-content/uploads/2018/08/AWG-Newsletter-No2-2010.pdf	
AWG	Newsletter	2011	http://quaternary.stratigraphy.org/wp-content/uploads/2018/08/Anthropocene-Working-Group-Newsletter-3A.pdf	
AWG	Newsletter	2012	http://quaternary.stratigraphy.org/wp-content/uploads/2018/08/Anthropocene-Working-Group-Newsletter-Vol-4-Final.pdf	
AWG	Newsletter	2013- Sep. 2014	http://quaternary.stratigraphy.org/wp-content/uploads/2018/08/Anthropocene-Working-Group-Newsletter-Vol-5.pdf	
AWG	Newsletter	Oct. 2014-2015	http://quaternary.stratigraphy.org/wp-content/uploads/2018/08/Anthropocene-Working-Group-Newsletter-Vol-6-release.pdf	
AWG	Newsletter	2016-2017	http://quaternary.stratigraphy.org/wp-content/uploads/2018/08/Anthropocene-Working-Group-Newsletter-Vol-7-release.pdf	
AWG	Newsletter	2018	http://quaternary.stratigraphy.org/wp-content/uploads/2018/12/Anthropocene-Working-Group-Newsletter-Vol-8.pdf	
AWG	Newsletter	2019	http://quaternary.stratigraphy.org/wp-content/uploads/2020/09/Anthropocene-Working-Group-Newsletter-Vol-9-final.pdf	
AWG	Newsletter	2020	http://quaternary.stratigraphy.org/wp-content/uploads/2021/03/AWG-Newsletter-2020-Vol-10.pdf	
AWG	Newsletter	2021	http://quaternary.stratigraphy.org/wp-content/uploads/2022/03/AWG-Newsletter-2021_final.pdf	
AWG	Newsletter	2022	http://quaternary.stratigraphy.org/wp-content/uploads/2023/03/AWG-2022-Newsletter_Final.pdf	
AWG	Newsletter	2023	http://quaternary.stratigraphy.org/wp-content/uploads/2024/03/AWG-2023-Newsletter_final_CWST_16-March.pdf	

Table A2 lists documents and reports issued by the Subcommission on Quaternary Stratification (SQS), the International Commission on Stratigraphy (ICS) and the International Union of Geological Sciences (IUGS) reports appear annually with exceptions. The original specific and master locations are given. See table A1 for updated link for master location for the site "quaternary.stratigraphy.org"

Organisation	Document Type	Reference Period	Specific Location (hyperlink)	Master Location (hyperlink)
SQS	AWG Binding Vote	2019	http://quaternary.stratigraphy.org/working-groups/anthropocene/	http://quaternary.stratigraphy.org
SQS	Re-establishing SQS	2002	http://quaternary.stratigraphy.org/annual-reports/ar2002/	http://quaternary.stratigraphy.org/blog/; (search: 2002)
ICS	Business Meeting Report	25 Feb 2022	https://stratigraphy.org/news/139_ICSBusinessMeetingMinutes.pdf	https://stratigraphy.org/news/139 ; https://stratigraphy.org
ICS	Business Meeting Report	31 Aug 2016	https://stratigraphy.org/files/ICS-Business-IGC35.pdf	https://stratigraphy.org/archives
ICS	Report Holocene Subdivision	July 2018	http://quaternary.stratigraphy.org/2018/07/	http://quaternary.stratigraphy.org/blog/
ICS	Statute	n.a.	https://stratigraphy.org/statutes	https://stratigraphy.org
ICS	Annual Report	2009 to 2022	https://stratigraphy.org/files/ICS_AnnReport2009.pdf https://stratigraphy.org/files/ICS_AnnReport2010.pdf https://stratigraphy.org/files/ICS_AnnReport2011.pdf https://stratigraphy.org/files/ICS_AnnReport2012.pdf https://stratigraphy.org/files/ICS_AnnReport2013.pdf https://stratigraphy.org/files/ICS_AnnReport2014.pdf https://stratigraphy.org/files/ICS_AnnReport2015.pdf https://stratigraphy.org/files/ICS_AnnReport2016.pdf https://stratigraphy.org/files/ICS_AnnReport2017.pdf https://stratigraphy.org/files/ICS_AnnReport2018.pdf https://stratigraphy.org/files/ICS_AnnReport2019.pdf https://stratigraphy.org/files/ICS_AnnReport2020.pdf https://stratigraphy.org/files/ICS_AnnReport2021.pdf https://stratigraphy.org/files/ICS_AnnReport2022.pdf	https://stratigraphy.org/archives

(continued)

Table A2 (continued)

Organisation	Document Type	Reference Period	Specific Location (hyperlink)	Master Location (hyperlink)
ICS	Decision Report Quaternary	28 Sep 2005	https://stratigraphy.org/files/Q1.pdf	https://stratigraphy.org/archives
ICS	Subcommission Report	2009 to 2015	https://stratigraphy.org/files/ICS_SubcommReport2010.pdf https://stratigraphy.org/files/ICS_SubcommReport2010.pdf https://stratigraphy.org/files/ICS_SubcommReport2011.pdf https://stratigraphy.org/files/ICS_SubcommReport2012.pdf https://stratigraphy.org/files/ICS_SubcommReport2013.pdf https://stratigraphy.org/files/ICS_SubcommReport2014.pdf https://stratigraphy.org/files/ICS_SubcommReport2015.pdf [gap] https://stratigraphy.org/files/ICS_SubcommReport2022.pdf	https://stratigraphy.org/archives
IUGS	Statutes & Bylaws	n.a.	https://www.iugs.org/statutes-bylaws	https://www.iugs.org
IUGS	Annual Reports	2001-2006, 2008-2012, 2013/14, 2015-2022	https://www.iugs.org/annual-reports	https://www.iugs.org/documents

Table A3 shows the Anthropocene Working Group (AWG) membership as reported in the AWG's newsletter. The AWG members are listed chronologically, i.e., the year the membership started. The symbols denote: "/" period prior to membership; "+" member, "--", no member, and "+\-" left AWG in dissent with the AWG's proposal. The last column lists the status of the AWG member (voting rights, advisory) and whether the member has died

Members of the Anthropocene Working Group	2009	2010	2011	2012	2013_14	2014_15	2016_17	2018	2019	2020	2021	2022	2023	Voting Rights (proposal) or advisor
Paul Crutzen	+	+	+	+	+	+	+	+	+	+	–	–	–	honorary (dec.)
Erle Ellis	+	+	+	+	+	+	+	+	+	+	+	+	+\-	2020: Adv
Mike Ellis	+	+	+	+	+	+	+	+	+	–	–	–	–	–
Philip Gibbard	+	+	+	+	+	+	+	+	+	+	+	+	+\-	2020: V
Alan Haywood	+	+	+	+	+	+	–	–	–	–	–	–	–	–
Andrew Kerr	+	+	+	+	–	–	–	–	–	–	–	–	–	–
Carlos Nobre	+	+	+	+	–	–	–	–	–	–	–	–	–	–
Eric Odada	+	+	+	+	+	+	+	+	+	+	+	+	+	2020: Adv
Simon Price	+	+	+	+	+	+	–	–	–	–	–	–	–	–
Mary Scholes	+	+	+	+	+	+	–	–	–	–	–	–	–	–
Will Steffen	+	+	+	+	+	+	+	+	+	+	+	–	–	2021: Adv (dec.)
Davor Vidas	+	+	+	+	+	+	+	+	+	+	+	+	+	2020: Adv
Mike Walker	+	+	+	+	–	–	–	–	–	–	–	–	–	–
Mark Williams	+	+	+	+	+	+	+	+	+	+	+	+	+	2020: V
Jan Zalasiewicz	+	+	+	+	+	+	+	+	+	+	+	+	+	2020: V
An Zhisheng	+	+	+	+	+	+	+	+	+	+	+	+	+	2020: Adv
Alejandro Cearreta	/	/	+	+	+	+	+	+	+	+	+	+	+	2020: V
Jacques Grinevald	/	/	+	+	+	+	+	+	+	+	+	+	+	2020: Adv

(continued)

Table A3 (continued)

Members of the Anthropocene Working Group	2009	2010	2011	2012	2013_14	2014_15	2016_17	2018	2019	2020	2021	2022	2023	Voting Rights (proposal) or advisor
Andrew Revkin	/	/	+	+	+	+	–	–	–	–	–	–	–	–
Ian Fairchild	/	/	/	+	+	+	+	+	+	+	+	+	+	2020: V
John McNeill	/	/	/	+	+	+	+	+	+	+	+	+	+	2020: Adv
Clément Poirier	/	/	/	+	+	+	+	+	+	+	+	+	+	2020: Adv
Dan Richter	/	/	/	+	+	+	+	+	+	+	+	+	+	2020: Adv
Colin Waters	/	/	/	+	+	+	+	+	+	+	+	+	+	2020: V
Tony Barnosky	/	/	/	/	+	+	+	+	+	+	+	+	+	2020: V
Matt Edgeworth	/	/	/	/	+	+	+	+	+	+	+	+	+\-	2020: Adv
Peter Haff	/	/	/	/	+	+	+	+	+	+	+	+	–	2021: Adv (dec.)
Reinhold Leinfelder	/	/	/	/	+	+	+	+	+	+	+	+	+	2020: V
Bruce Smith	/	/	/	/	+	–	–	–	–	–	–	–	–	–
Michael Wagreich	/	/	/	/	+	+	+	+	+	+	+	+	+	2021: V
Alex Wolfe	/	/	/	/	+	+	+	+	+	–	–	–	–	–
Agnieszka Gałuszka	/	/	/	/	/	+	+	+	+	+	+	+	+	2020: V
Irka Hajdas	/	/	/	/	/	+	+	+	+	+	+	+	+	2020: V
Cath Neal	/	/	/	/	/	+	+	+	+	–	–	–	–	–
Victoria C. Smith	/	/	/	/	/	+	–	–	–	–	–	–	–	–
Colin Summerhayes	/	/	/	/	/	+	+	+	+	+	+	+	+	2020: V

Jaia Syvitski	/	/	/	/	/	+	+	+	+	+	+	+	+	2020: V
Scott Wing	/	/	/	/	/	+	+	+	+	+	+	+	+	2020: V
Juliana Assunção Ivar do Sul	/	/	/	/	/	/	+	+	+	+	+	+	+	2020: Adv
Catherine Jeandel	/	/	/	/	/	/	+	+	+	+	+	+	+	2020: Adv
Naomi Oreskes	/	/	/	/	/	/	+	+	+	+	+	+	+	2020: Adv
Martin Head	/	/	/	/	/	/	/	+	+	+	+	+	+	2020: V
Neil Rose	/	/	/	/	/	/	/	/	+	+	+	+	+	2020: V
Bill Shotyk	/	/	/	/	/	/	/	/	+	+	+	+	+	2020: Adv
Andy Cundy	/	/	/	/	/	/	/	/	/	/	+	+	+	2020: V
Han Yongming	/	/	/	/	/	/	/	/	/	/	+	+	+	2020: V
Francine McCarthy	/	/	/	/	/	/	/	/	/	/	+	+	+	2020: V
Yoshiki Saito	/	/	/	/	/	/	/	/	/	/	+	+	+	2020: V
Simon Turner	/	/	/	/	/	/	/	/	/	/	+	+	+	2020: V
Jens Zinke	/	/	/	/	/	/	/	/	/	/	+	+	+	2020: V
Barbara Fiałkiewicz-Kozieł	/	/	/	/	/	/	/	/	/	/	+	+	+	2021: V

The manufacturer's authorised representative in the EU is Springer Nature Customer Service Centre GmbH, Europaplatz 3, 69115 Heidelberg, Germany. If you have any concerns regarding our products, please contact ProductSafety@springernature.com

Printed and bound by CPI Group (UK) Ltd, Croydon, CR0 4YY
30/09/2026
02225297-0007